SIMPLY
THE BRAIN

Penguin Random House

Editors Michael Clark, Joanna Edwards,
Miezan van Zyl
Senior US Editor Kayla Dugger
Senior Designer Ali Scrivens
Assistant Editor Emily Kho
Managing Editor Angeles Gavira
Managing Art Editor Michael Duffy
Production Editor Kavita Varma
Senior Production Controller
Meskerem Berhane
Jacket Design Development Manager
Sophia M.T.T.
Jacket Designer Akiko Kato
Associate Publishing Director Liz Wheeler
Art Director Karen Self
Publishing Director Jonathan Metcalf

First American Edition, 2022
Published in the United States by DK Publishing
1745 Broadway, 20th Floor, New York, NY 10019

A catalog record for this book
is available from the Library of Congress.
ISBN 978-0-7440-6011-9

Printed and bound in China

For the curious
www.dk.com

CONSULTANT

Rita Carter is the author of many books about the human brain, including DK's *The Brain Book*. She holds a doctorate in brain science from Leuven University, Belgium.

CONTRIBUTORS

Liam Drew is a science writer and journalist. A former neurobiologist, he now writes feature stories and books about neuroscience, biomedical research, and evolutionary biology.

Alina Ivan works in mental health research at the Institute of Psychiatry, Psychology & Neuroscience (IoPPN), King's College London. She is also a science presenter and develops partnerships for the mental health platform Augmentive. She holds degrees from the University of Exeter.

Susan Watt is a science writer and editor. She holds postgraduate degrees in psychology and philosophy and has written widely about psychology research, with special interests in reasoning and ADHD.

Emma Yhnell is an award–winning science communicator and a senior lecturer at Cardiff University in Wales. She loves teaching and has published research into Huntington's disease as well as blogs and books for public audiences.

CONTENTS

7   **INTRODUCTION**

THE **HUMAN BRAIN**

10   **WHAT IS A BRAIN?**
     The human brain

12   **TWO HALVES**
     The cerebral hemispheres

14   **THREE MAIN PARTS**
     Regions of the brain

15   **COMPLEX CORTEX**
     The cerebral cortex

16   **GENERATING RESPONSES**
     The limbic system

17   **SURVIVAL MACHINE**
     Ensuring safety

18   **BRAIN CELLS**
     Neurons

19   **MAKING CONNECTIONS**
     Synapses

20   **PASSING MESSAGES**
     Signals to the brain

22   **TYPES OF BRAIN WAVES**
     Neural activity

23   **TRAINING THE BRAIN**
     Neurofeedback

24   **THE ADAPTABLE BRAIN**
     Neural plasticity

26   **MAPPING THE BRAIN**
     MRI scans

27   **TRACKING ACTIVITY**
     Magnetoencephalography

THE **SENSATIONAL WORLD**

30   **SENSING THE WORLD**
     The senses

32   **VISUAL PATHWAYS**
     Seeing

33   **FALSE PERCEPTIONS**
     Hallucinations

34   **GOOD VIBRATIONS**
     Hearing

35   **COCKTAIL PARTY EFFECT**
     Filtering sounds

36   **UNDER PRESSURE**
     Touching

37   **DETECTING FLAVOR**
     Tasting

38   **ON THE SCENT**
     Smelling

39   **BLENDED SENSES**
     Synesthesia

MOODS AND **EMOTIONS**

42   **HOW DO YOU FEEL?**
     What are emotions?

43   **EMOTIONAL RANGE**
     Types of emotion

44   **IN A MOOD**
     What are moods?

46 **GENERATING EMOTIONS**
Emotions and the brain

47 **HIDDEN FEELINGS**
Unconscious emotions

48 **FEELGOODS**
Positive emotions

49 **FEELBADS**
Negative emotions

50 **ARTISTIC COMMUNICATION**
Visual art and reflection

52 **MUSIC AND THE BRAIN**
The effects of music

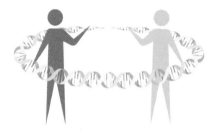

INDIVIDUALITY

56 **NATURE VS. NURTURE**
Genetics and the environment

58 **THE CHANGING BRAIN**
Development

60 **MORE ALIKE THAN DIFFERENT**
Male and female brains

61 **GENETIC BLUEPRINT**
Genes

62 **NEW CONNECTIONS**
Molded by experience

63 **EVOLVING CHARACTERISTICS**
Personality

64 **WHO ARE YOU?**
Personality types

MEMORY AND LEARNING

68 **TYPES OF MEMORY**
Memory categories

70 **STORING MEMORIES**
Storage

71 **RECALLING MEMORIES**
Recall

72 **IDENTIFYING THE FAMILIAR**
Recognition memory

73 **WHY WE FORGET**
Memory loss

74 **IMPROVING YOUR MEMORY**
Memory aids

75 **FALSE MEMORIES**
Memory distortions

76 **MEMORIES THAT AREN'T FORGOTTEN**
Habits

COMMUNICATION

80 **THE LANGUAGE BLUEPRINT**
Is language hardwired?

81 **TAKING IT ALL IN**
Understanding

82 **VERBAL AND NONVERBAL**
Speaking and sign language

84 **SPEAKING WITHOUT WORDS**
Body language

86 **THE WRITTEN WORD**
Reading and writing

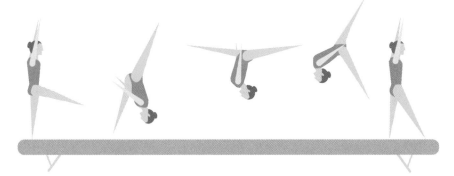

THE **BRAIN** AND **ACTION**

90 **ORGANIZED MIND**
Body maps

91 **PHANTOM SENSATIONS**
Phantom limbs

92 **DISTORTED REFLECTION**
Dysmorphia

93 **CONTROL AREAS**
Cognitive network

94 **MAKING A MOVE**
Action planning

96 **KNOWING YOUR PLACE**
Proprioception and kinesthesia

98 **COORDINATION AND FINE-TUNING**
Gross and fine motor skills

THE **SOCIAL** BRAIN

102 **DO I KNOW YOUR FACE?**
Facial recognition

104 **THE OTHER'S PERSPECTIVE**
Theory of mind

106 **READING OTHERS**
Mind reading

107 **MONKEY SEE, MONKEY DO**
Mirror neurons

108 **FEELING WITH**
Empathy

109 **A DEEP CONNECTION**
Bonding

110 **ARE WE JUST SHEEP?**
The herd instinct

112 **THIS FEELS WRONG**
Morality

113 **ALL BY MYSELF**
Loneliness

114 **LOVE ON THE BRAIN**
Romantic love

115 **BORN THIS WAY**
Gender and sexuality

NEURODIVERSITY

118 **WE'RE NOT ALL THE SAME**
Neurodiversity

120 **DIFFICULT PATTERNS**
Personality disorders

122 **BRILLIANT MINDS**
Genius

123 **A MISSING CONSCIENCE**
Psychopaths

124 **A LOSS OF REALITY**
Schizophrenia

126 **A VARIED SPECTRUM**
Autism

THE RATIONAL BRAIN

THE CONSCIOUS BRAIN

130 **INTUITION OR REASON?**
What is rational?

131 **SPLIT-SECOND PROCESSING**
Gathering evidence

132 **WHAT GRABS YOU?**
Editing the facts

133 **DOUBLE-EDGED THINKING**
Judging and deciding

134 **ISSUING ORDERS**
Taking action

135 **CHOOSING A PATH**
Making decisions

136 **THE BIRTH OF IDEAS**
Creativity

138 **A NEW STATE OF MIND**
Hypnosis

139 **QUIETING THE BRAIN**
Meditation

142 **WHAT IS CONSCIOUSNESS?**
Consciousness

143 **UNLOCK THE UNCONSCIOUS**
Unconsciousness

144 **FINDING CONSCIOUSNESS**
The science of consciousness

146 **ARTIFICIAL INTELLIGENCE**
Machine consciousness

147 **FRAMES OF CONSCIOUSNESS**
Altered states

148 **TIME AND THE BRAIN**
Body clocks

149 **GRABBING YOUR ATTENTION**
Attention

150 **NECESSARY REST**
Sleep

151 **TYPES OF SLEEP**
Sleep cycles

152 **STRUGGLES WITH SLEEP**
Sleep disorders

153 **DREAM WORLDS**
Dreaming

154 **THE DREAMING MIND**
Dream consciousness

155 **FALSE AWAKENING**
Lucid dreaming

156 Index

160 Acknowledgments

THE BRAIN

We can never discover everything about the human brain because if it was simple enough to understand, we would be too simple even to try. One of the things the brain does best, though, is exploration, and the curiosity and ingenuity of the human brain have led us to find out its basic structure and many of the processes that occur in it. This knowledge explains much about the way we think, feel, and behave.

Our brain evolved to keep our body alive, and everything it does stems from that. Our senses, for example, tell us what is happening around us and our motor abilities allow us to react to it. Pleasure and pain tell us what is beneficial and what is harmful, and memory ensures we carry that information as a guide to future action—a painful burn is very effective at preventing us from getting too close to fire again.

Many other animals can do all this, but human brains can do far more. In the 8 or 9 million years since we diverged from our primate ancestors, our brain has tripled in size, allowing it to process information in a very complex way. It has also grown a multilayered skin, the neocortex, which folds over in front to form frontal lobes that monitor and mediate activity occurring in the rest of the brain. These unique features endow us with exceptional skills: abstract thinking that allows us to imagine worlds beyond our senses; literacy, which lets us pass on knowledge through generations; and imagination, which gives us the ability to create our own future.

The arrival of these sophisticated skills did not obliterate the old drives, however, and our brain still retains the parts that produce emotional reactions. In fact, emotion still drives most of our behavior, but some of the brain processes that evolved to keep us safe do not work so well in today's world. Rage once helped us see off territorial challengers, for example, but it is disastrous when we're driving a car. Our newly evolved "rational" brain—the neocortex—has to work hard to keep extreme emotions in check.

THE H
BRAI

U M A N
N

The human brain contains billions of electrically powered cells called neurons. They are connected by long tendrils (axons) forming a super-dense network of neural pathways. Signals flow through this network in ever-changing patterns, creating our thoughts, sensations, and emotions. Some neural pathways are genetically determined, while others are formed by experience. The brain divides naturally into two mirrored hemispheres, which are further divided into four lobes. Each lobe contains thousands of distinct but interactive regions, or modules. The brain merges with the spinal cord, which carries nerve fibers to the farthest reaches of the body.

WHAT IS A BRAIN?

Nerves evolved to control the behavior of animals in order to maximize their survival and reproductive potential. In some animals, nerves compacted into a central, integrated organ called a brain, which forms part of the central nervous system (CNS). This comprises the brain and spinal cord, connected by the brainstem. Other animals have a distributed nervous system; for example, the octopus has separate control systems in each limb. The human brain is unusual for its highly developed frontal lobe, which allows us to generate a rich perceptual world and to achieve great feats of thought and imagination. Physical action seems to be the least of the brain's abilities, but most of the organ is still given over to controlling our bodies.

Weight:
On average, the brain of an adult human weighs 2.6–3.0 lb (1.2–1.35 kg). It is about 73 percent water.

Volume:
The average volume ranges from 70 to 78 cubic in (1,150 to 1,275 cubic cm), but this decreases with age.

Gray matter:
About 40 percent of the brain's tissue is gray matter (tightly packed nerve-cell bodies).

White matter:
Long extensions of nerve cells covered in sheaths of fat are about 60 percent of brain tissue.

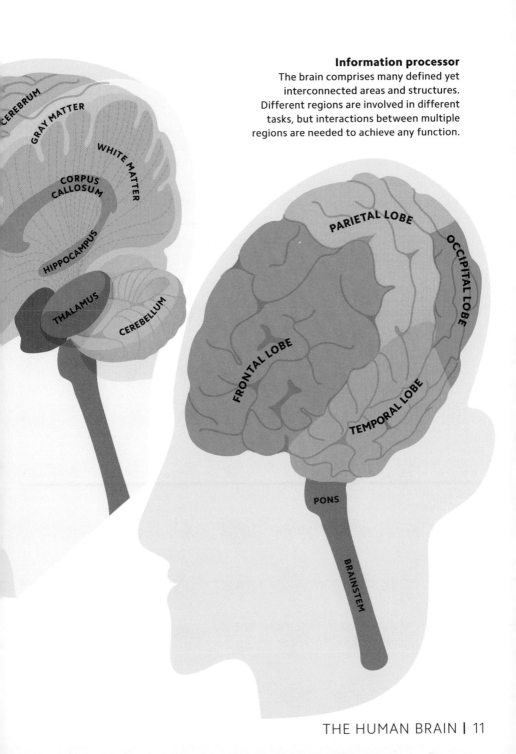

Information processor
The brain comprises many defined yet interconnected areas and structures. Different regions are involved in different tasks, but interactions between multiple regions are needed to achieve any function.

CEREBRUM

GRAY MATTER

WHITE MATTER

CORPUS CALLOSUM

HIPPOCAMPUS

THALAMUS

CEREBELLUM

PARIETAL LOBE

OCCIPITAL LOBE

FRONTAL LOBE

TEMPORAL LOBE

PONS

BRAINSTEM

TWO HALVES

The brain has two mirror-image hemispheres. The old notion that our creative abilities arise from the right-hand side and our colder analytical skills from the left and that people show a dominance of one hemisphere over another is now debunked. However, the two hemispheres do have significant functional differences. Language, for instance, relies on the left hemisphere in most people, and the right is crucial for abstract reasoning. For still-unknown reasons, each side of the brain receives sensory input from the opposite side of the body and controls the opposite side, too. Nearly everyone favors one more dexterous hand over the other—likewise, one of their feet.

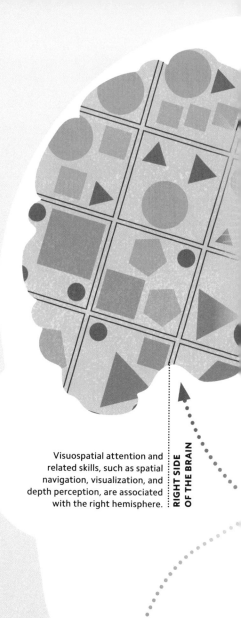

Visuospatial attention and related skills, such as spatial navigation, visualization, and depth perception, are associated with the right hemisphere.

RIGHT SIDE OF THE BRAIN

Right side of the body
Around 90 percent of people are right-handed, meaning they prefer to use this hand for performing tasks requiring dexterity and in general. But only around 60 percent are right-footed; 30 percent of people are "mixed footed."

**LEFT SIDE
OF THE BRAIN**

In most people, language
generation, articulation,
and word comprehension
happens in the brain's
left hemisphere.

Left side of the body
No one is completely sure why
some people are left-handed—
attempts to finds genes for
this trait have failed. However,
babies in the womb already
use one hand more than the
other, sometimes sucking the
thumb of their favored hand.

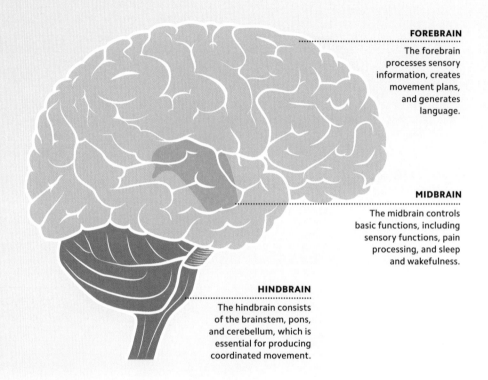

FOREBRAIN

The forebrain processes sensory information, creates movement plans, and generates language.

MIDBRAIN

The midbrain controls basic functions, including sensory functions, pain processing, and sleep and wakefulness.

HINDBRAIN

The hindbrain consists of the brainstem, pons, and cerebellum, which is essential for producing coordinated movement.

THREE MAIN PARTS

The brain is divided into three unequally sized parts, known as the forebrain, midbrain, and hindbrain. These divisions are based on how the brain develops in the womb, and each part has distinctive functions. The forebrain is massively expanded in humans—comprising almost 90 percent of the brain's weight—and includes the cerebral cortex, the brain's folded outer layer. The midbrain is a small central part. The hindbrain includes the cerebellum, which protrudes from the back of the brain.

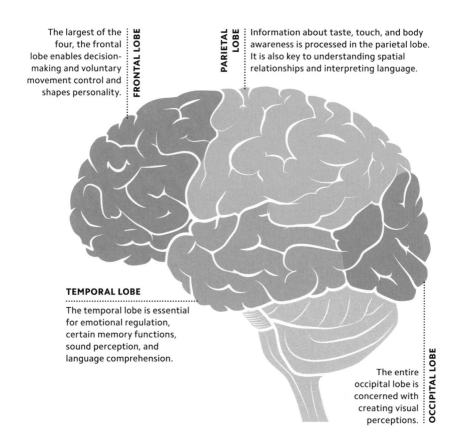

FRONTAL LOBE

The largest of the four, the frontal lobe enables decision-making and voluntary movement control and shapes personality.

PARIETAL LOBE

Information about taste, touch, and body awareness is processed in the parietal lobe. It is also key to understanding spatial relationships and interpreting language.

TEMPORAL LOBE

The temporal lobe is essential for emotional regulation, certain memory functions, sound perception, and language comprehension.

OCCIPITAL LOBE

The entire occipital lobe is concerned with creating visual perceptions.

COMPLEX CORTEX

The cerebral cortex is the largest part of the human brain. Its multiple folds allow more cortical tissue to fit inside the skull. The cortex carries out the brain's most complex information processing, and its massive expansion in humans is elemental to the sophistication of human intelligence. Divisible into over 50 regions, each with a unique internal organization and different connections to other brain regions, the cortex is conventionally divided into four main lobes. Each lobe has characteristic primary functions.

GENERATING RESPONSES

The limbic system is a collection of structures that lie in the brain's interior. It is elemental to generating and controlling our emotional lives and for forming memories. The powerful interplay between smells, memories, and feelings arises here. The limbic structures are highly connected to the cortex and the brain's reward centers and control our motivational states and goal-directed behaviors. To help do this, the limbic system also has a large influence on the body's hormonal system.

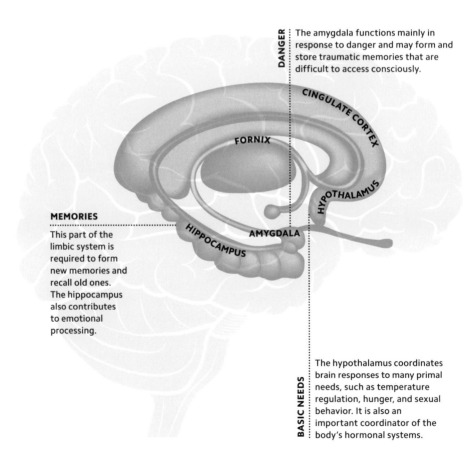

DANGER
The amygdala functions mainly in response to danger and may form and store traumatic memories that are difficult to access consciously.

CINGULATE CORTEX

FORNIX

HYPOTHALAMUS

MEMORIES
This part of the limbic system is required to form new memories and recall old ones. The hippocampus also contributes to emotional processing.

HIPPOCAMPUS

AMYGDALA

BASIC NEEDS
The hypothalamus coordinates brain responses to many primal needs, such as temperature regulation, hunger, and sexual behavior. It is also an important coordinator of the body's hormonal systems.

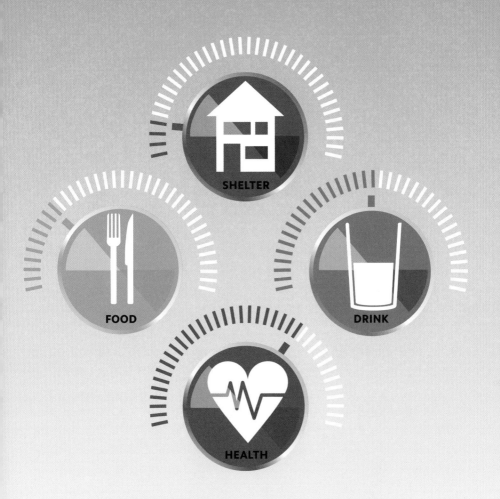

SURVIVAL MACHINE

The brain must monitor a person's most basic needs: Am I hungry or thirsty? Am I full? Am I too hot or too cold? Is there enough oxygen in my blood? Do I need to use the bathroom? Once these basic physiological needs are recognized, the brain instigates a behavioral response to seek food, water, or shelter. It also induces unconscious physiological responses, such as directing the body to breathe deeper, to sweat, or to shiver. These coordinated responses act to avert bodily harm.

BRAIN CELLS

Neurons are the cells in the brain that relay and process information. There are around 86 billion in a human brain. Neurons have a cell body from which wiry processes extend to form connections with other neurons. On one side, neurons have short dendrites, which collect incoming signals. On the other, a neuron has a very long axon, along which outgoing signals travel.

Neuronal pathways
Connecting neurons form vast networks that allow the brain to process many pieces of information at the same time.

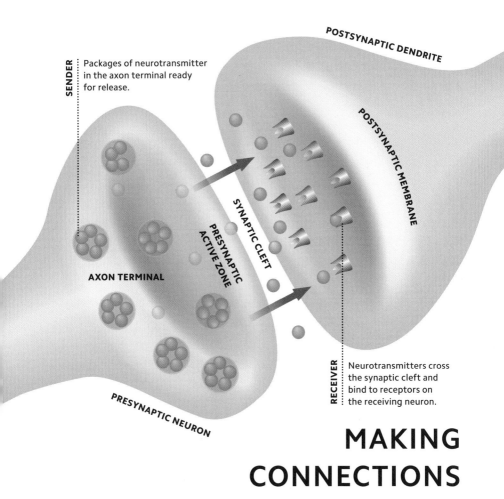

Packages of neurotransmitter in the axon terminal ready for release.

POSTSYNAPTIC DENDRITE

POSTSYNAPTIC MEMBRANE

SYNAPTIC CLEFT

PRESYNAPTIC ACTIVE ZONE

AXON TERMINAL

PRESYNAPTIC NEURON

RECEIVER Neurotransmitters cross the synaptic cleft and bind to receptors on the receiving neuron.

MAKING CONNECTIONS

The junction between two neurons is called a synapse. Neurons are separated by a synaptic cleft—a gap 10,000 times thinner than a human hair. An electrical impulse in the first (presynaptic) neuron causes the release of chemical messengers, known as neurotransmitters, which cross the synaptic cleft to the second (postsynaptic) neuron. They attach to receptors and change the electrical activity of that neuron, thus passing on the message.

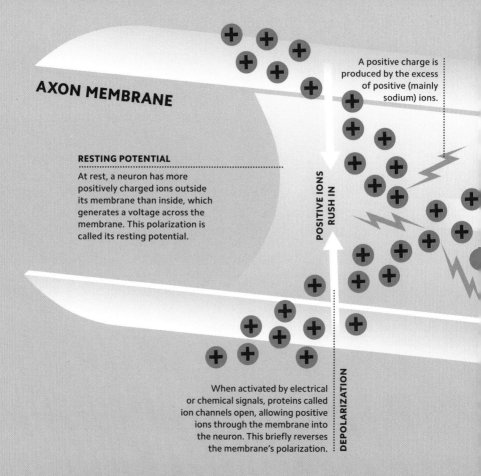

AXON MEMBRANE

A positive charge is produced by the excess of positive (mainly sodium) ions.

RESTING POTENTIAL

At rest, a neuron has more positively charged ions outside its membrane than inside, which generates a voltage across the membrane. This polarization is called its resting potential.

POSITIVE IONS RUSH IN

When activated by electrical or chemical signals, proteins called ion channels open, allowing positive ions through the membrane into the neuron. This briefly reverses the membrane's polarization.

DEPOLARIZATION

PASSING MESSAGES

Information travels around the brain via electrical pathways formed by axons. Each neuron has one of these long tendrils extending from it. Changes in the voltage across the surface of a neuron's dendrites and cell body generate brief electrical spikes—called action potentials—that move down the axon. Then, when action potentials reach the axon's outgoing synapses, they cause the release of chemical neurotransmitters. These diffuse to the receiving neuron, binding to receptors that change the voltage across that neuron's dendrites. A neuron's combined synaptic inputs make it more or less likely, in turn, to fire action potentials.

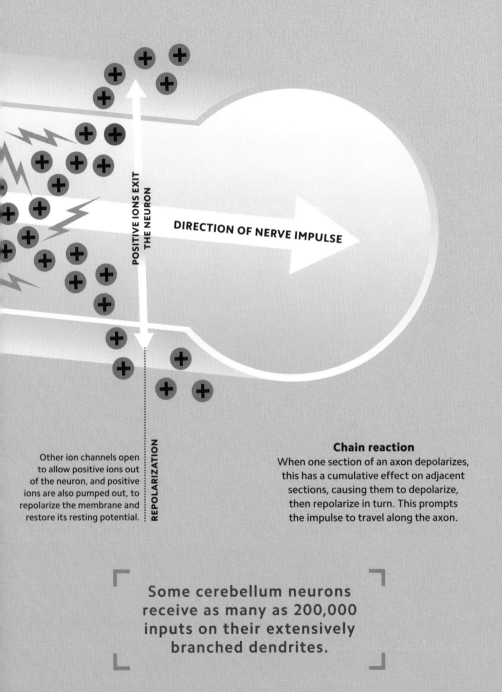

POSITIVE IONS EXIT THE NEURON

DIRECTION OF NERVE IMPULSE

REPOLARIZATION

Other ion channels open to allow positive ions out of the neuron, and positive ions are also pumped out, to repolarize the membrane and restore its resting potential.

Chain reaction
When one section of an axon depolarizes, this has a cumulative effect on adjacent sections, causing them to depolarize, then repolarize in turn. This prompts the impulse to travel along the axon.

Some cerebellum neurons receive as many as 200,000 inputs on their extensively branched dendrites.

TYPES OF BRAIN WAVES

When neurologists place electrodes on a person's scalp to record an electroencephalogram (EEG), they observe brain waves. Brain waves are oscillating changes in voltage that reflect the electrical activity of neurons. Five main waves are recognized, each with a distinct average frequency and an associated mental state; for example, the slowest, Delta, is associated with deep sleep. They appear as waves because although neurons can fire at any time, they are statistically more likely to fire spikes in a loosely synchronized fashion.

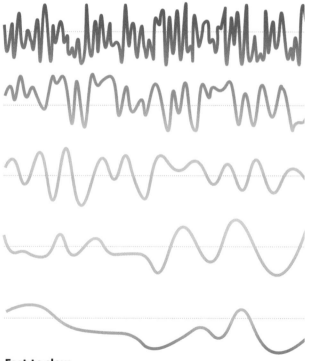

GAMMA
Dense Gamma waves are linked to learning and problem-solving tasks.

BETA
These high-frequency waves occur when one is awake and conscious.

ALPHA
These waves occur when one is calm but still alert and aware.

THETA
These waves indicate a state of deep relaxation or meditation.

DELTA
Low-frequency Delta waves are observed during deep sleep.

Fast to slow
Brain waves are measured in cycles per second, or hertz (Hz), and there are five main patterns.

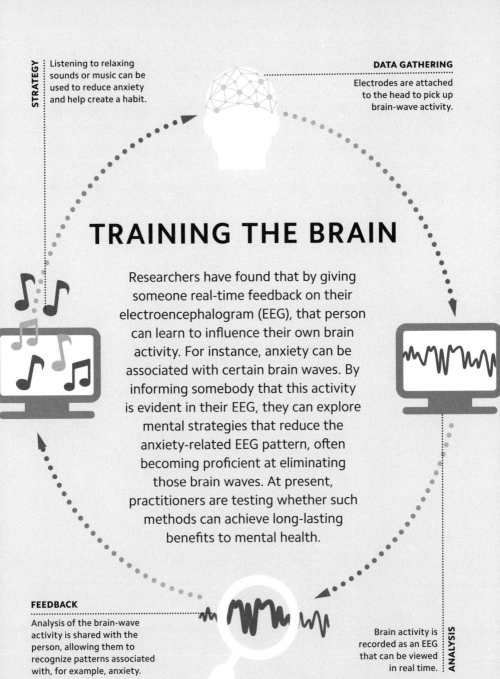

Listening to relaxing
sounds or music can be
used to reduce anxiety
and help create a habit.

DATA GATHERING

Electrodes are attached
to the head to pick up
brain-wave activity.

TRAINING THE BRAIN

Researchers have found that by giving
someone real-time feedback on their
electroencephalogram (EEG), that person
can learn to influence their own brain
activity. For instance, anxiety can be
associated with certain brain waves. By
informing somebody that this activity
is evident in their EEG, they can explore
mental strategies that reduce the
anxiety-related EEG pattern, often
becoming proficient at eliminating
those brain waves. At present,
practitioners are testing whether such
methods can achieve long-lasting
benefits to mental health.

FEEDBACK

Analysis of the brain-wave
activity is shared with the
person, allowing them to
recognize patterns associated
with, for example, anxiety.

Brain activity is
recorded as an EEG
that can be viewed
in real time.

ANALYSIS

THE ADAPTABLE BRAIN

Brains are often described as being "plastic," meaning that they change their form and function in response to experience. Such modifications underlie memory formation and learning, as well as the brain's ability to overcome some injuries or disease-related damage. There are two main forms of neural plasticity. With synaptic plasticity, the signal a synapse (see p.19) conveys becomes stronger or weaker in response to previous activation patterns. Structural plasticity is when dendrites and axons (see p.18) grow and retract, changing the connectivity patterns between neurons.

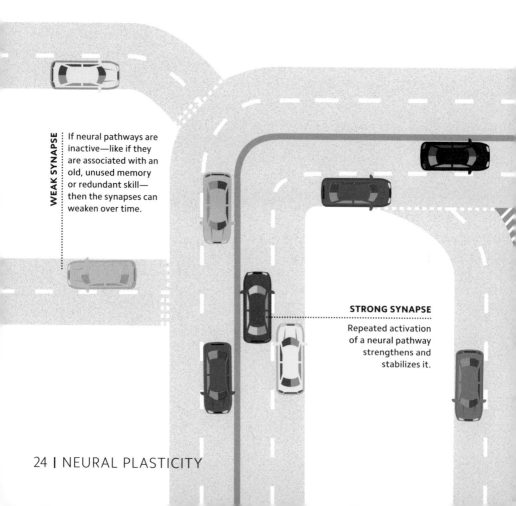

WEAK SYNAPSE

If neural pathways are inactive—like if they are associated with an old, unused memory or redundant skill—then the synapses can weaken over time.

STRONG SYNAPSE

Repeated activation of a neural pathway strengthens and stabilizes it.

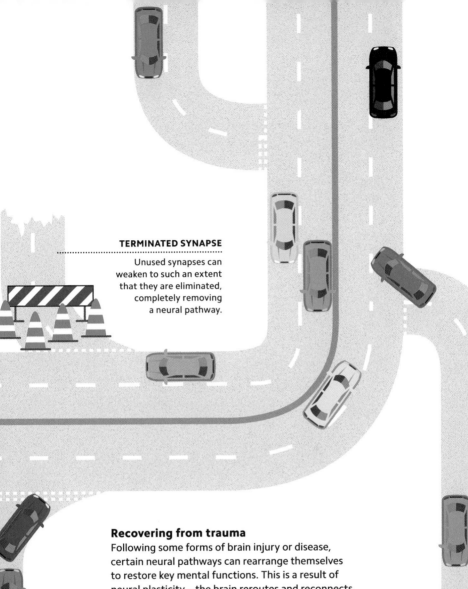

TERMINATED SYNAPSE
Unused synapses can weaken to such an extent that they are eliminated, completely removing a neural pathway.

Recovering from trauma

Following some forms of brain injury or disease, certain neural pathways can rearrange themselves to restore key mental functions. This is a result of neural plasticity—the brain reroutes and reconnects neural pathways so that neurons in undamaged areas carry out the functions previously executed by lost neurons. People who suffer strokes in brain areas associated with language often lose the ability to speak or write but can recover some functions when other areas of the brain are "rewired" to take them over.

MAPPING THE BRAIN

To study the living—healthy and unhealthy—human brain, scientists have developed several powerful brain-imaging technologies. The most widely used is magnetic resonance imaging (MRI), which uses strong magnetic fields and pulsed radio waves to reveal the anatomy of the brain. MRI scans can reveal structural changes or abnormalities, such as those caused by disease. A variant of this is functional MRI (fMRI), which maps blood flow in real time. Because active brain regions receive additional blood, fMRI scans can reveal which regions are active as a person performs certain mental tasks.

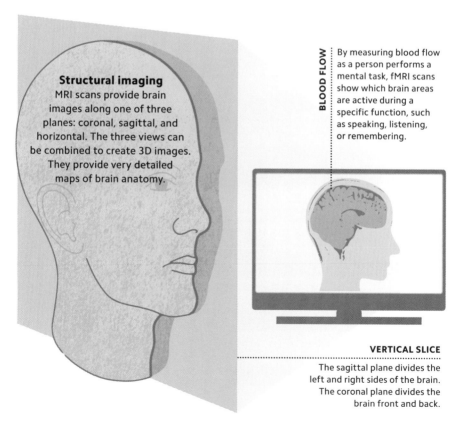

Structural imaging
MRI scans provide brain images along one of three planes: coronal, sagittal, and horizontal. The three views can be combined to create 3D images. They provide very detailed maps of brain anatomy.

BLOOD FLOW
By measuring blood flow as a person performs a mental task, fMRI scans show which brain areas are active during a specific function, such as speaking, listening, or remembering.

VERTICAL SLICE
The sagittal plane divides the left and right sides of the brain. The coronal plane divides the brain front and back.

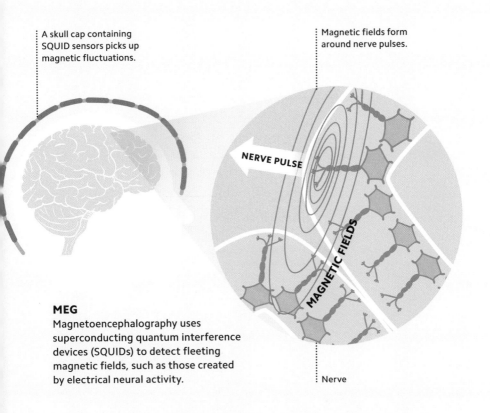

A skull cap containing SQUID sensors picks up magnetic fluctuations.

Magnetic fields form around nerve pulses.

NERVE PULSE

MAGNETIC FIELDS

MEG
Magnetoencephalography uses superconducting quantum interference devices (SQUIDs) to detect fleeting magnetic fields, such as those created by electrical neural activity.

Nerve

TRACKING ACTIVITY

Electrical impulses in the brain create extremely faint magnetic fields, which are constantly shifting with the neural activity. Scientists use magnetoencephalography (MEG)—a brain scan technique that detects rapid fluctuations in brain activity—to create images of the brain in action. The weakness of the magnetic fields created by neurons limits MEG's ability to pinpoint the location of the activity. However, MEG can be used to track the timing of the activity, which occurs in just a few thousandths of a second.

THE
SENSA
WORLD

TIONAL

Five basic senses (sight, hearing, smell, taste, and touch) are known to almost everyone, but humans have many more ways of sensing the world. Other senses include proprioception (the sense of where our body is in space), interioception (knowledge of what is happening inside our body), and chemoception (sensing chemicals, such as hormones). Some researchers claim that humans have more than 30 senses. Stimuli from the sense organs activate and are registered by a particular group of neurons called primary sense areas. These areas link to other areas of the brain where the stimuli are evaluated; identified; and, in some cases, made conscious.

SENSING THE WORLD

In an ever-changing world, brains constantly gather information about what is happening. Multiple sensory systems evolved to detect various types of signals, each relaying facets of our surroundings. In sensory organs, physical and chemical stimuli are converted into nerve impulses and these travel along nerve fibers to brain regions dedicated to processing individual senses. Sensory information is then broadcast across the brain to other brain regions, so the different sensory streams are integrated to create an overall perception of the world.

Smell

Humans can perceive a trillion different odors. Smell alone among the senses has a direct link to areas of the brain involved in emotions and memory, resulting in a powerful link between scents and memories.

See

Our eyes register light reflected from objects and transform it into electrical impulses that then travel to the occipital lobe at the back of the brain. We perceive objects in as little as half a second.

Taste

Two areas of the brain are devoted to taste. The first, in the frontal lobe, tells us what is good or bad. The second, between the temporal and frontal lobes, distinguishes the type of taste.

Hear

Our ears respond to vibrations of the air. These vibrations are converted into electrical impulses that are sent to the brain to analyze and interpret. Hearing is the most developed of the senses at birth.

Touch

Nerve receptors buried in the skin respond to stimulation from touch and temperature. Receptors in hair follicles are very sensitive to air movements, so we can detect things that have not touched our skin.

VISUAL PATHWAYS

Vision begins when light strikes our retina—an outgrowth of the brain at the back of each eye. Light reflected from objects stimulates an array of color-sensitive cone receptors and monochromatic rod receptors in the retina. Signals then travel along the optic nerve to the thalamus, where they are relayed to the visual cortex at the back of the brain. The visual cortex interprets these signals to create our visual perceptions. It contains an array of interlinked areas that process different aspects of the world, such as color and motion.

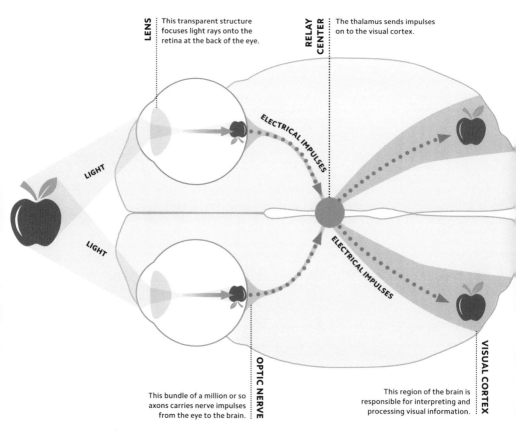

LENS
This transparent structure focuses light rays onto the retina at the back of the eye.

RELAY CENTER
The thalamus sends impulses on to the visual cortex.

ELECTRICAL IMPULSES

LIGHT

LIGHT

ELECTRICAL IMPULSES

OPTIC NERVE
This bundle of a million or so axons carries nerve impulses from the eye to the brain.

VISUAL CORTEX
This region of the brain is responsible for interpreting and processing visual information.

OLFACTORY

GUSTATORY

VISUAL

AUDITORY

TACTILE

FALSE PERCEPTIONS

Hallucinations are perceptions that have no basis in the external world. They may be sounds (typically voices); images (ghosts); or even tastes, touch, or smells. False perceptions occur when neural circuits in the brain fire even though they have not been stimulated by the sense organs. Imagination and recollection, which involve seeing something in our "mind's eye," are types of hallucinations, but normally the perceptions produced this way are too weak to be mistaken for external events.

GOOD VIBRATIONS

The ear turns sound waves, tiny fluctuations in air pressure, into electrical impulses that stimulate the auditory cortex. Different neurons in the auditory cortex responds to different frequencies; according to which ones are stimulated, we hear correspondingly different sounds. The position of the ears—on the side of the head—means that one ear usually receives the sound waves a split second before the other, and the brain uses that difference to determine where the sound comes from.

Pitch and tone

The faster a sound wave vibrates, the higher pitched the noise we perceive. The bigger the vibrations, the louder it sounds. Different hair cells in the cochlea respond to different sounds and pass on the pattern of the vibrations as nerve messages.

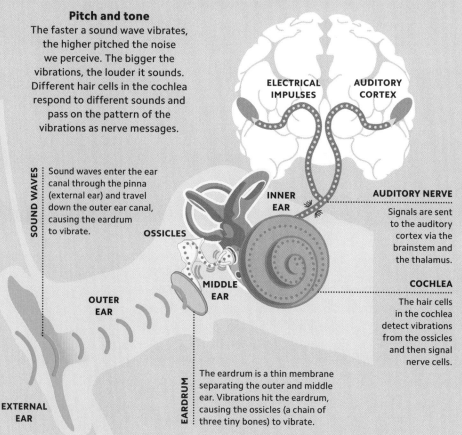

ELECTRICAL IMPULSES

AUDITORY CORTEX

SOUND WAVES

Sound waves enter the ear canal through the pinna (external ear) and travel down the outer ear canal, causing the eardrum to vibrate.

OSSICLES

INNER EAR

AUDITORY NERVE

Signals are sent to the auditory cortex via the brainstem and the thalamus.

MIDDLE EAR

OUTER EAR

COCHLEA

The hair cells in the cochlea detect vibrations from the ossicles and then signal nerve cells.

EXTERNAL EAR

EARDRUM

The eardrum is a thin membrane separating the outer and middle ear. Vibrations hit the eardrum, causing the ossicles (a chain of three tiny bones) to vibrate.

COCKTAIL PARTY EFFECT

Often, in a noisy world, only one particular source of sound is important to us. All sounds stimulate the ears equally, and neural signals relating to each enter the brain to be processed to some degree. Words that are important to us, however, get noted unconsciously, and this triggers brain processes related to attention to amplify them while dulling sounds that are not interesting.

Maria! Your brain registers important words and directs its attention.

Selective hearing
The ability to hear your own name without being distracted by loud background noise has become known as the "cocktail party effect."

LIGHT BREEZE	TEMPERATURE CHANGE	BRUSH OF A FEATHER	GENTLE TOUCH	FIRM PRESS	VIBRATION

ROOT HAIR PLEXUS	FREE NERVE ENDINGS	MERKEL'S DISKS	MEISSNER'S CORPUSCLES	RUFFINI ENDINGS	PACINIAN CORPUSCLE

TYPES OF SENSORY NERVE ENDING

UNDER PRESSURE

An array of sensory nerve endings lies beneath the surface of the skin, ready to detect forces pressing on the body. These nerve endings are densest in the fingertips and lips, where our sense of touch is most discriminating. There are also nerve endings sensitive to temperature, chemical irritants, and stimuli strong enough to damage the body, which we typically perceive as painful. Peripheral nerves start in the sensory cortex and run down through the brainstem and spinal cord to end in muscles, joints, and skin. Stimuli trigger the ends to send signals back up to the sensory cortex.

Sensory homunculus
A sensory homunculus is a body modeled in proportion to the area of the somatosensory cortex devoted to it. The lips, hands, and fingertips are enlarged on this distorted figure, as they send the most inputs to the brain.

DETECTING FLAVOR

Your tongue is covered in papillae, little lumps that contain taste buds made of nerve endings that respond to chemicals in food. When they are stimulated, they send signals to a deeply embedded part of the brain. This area distinguishes basic tastes: saltiness, umami (savory taste), sweetness, bitterness, and sourness. Subtle flavors are produced via a secondary route, which uses parts of the brain devoted to smell. Food molecules also waft into the nasal cavity and stimulate olfactory receptors, and these two senses combined give our perception of taste its full richness.

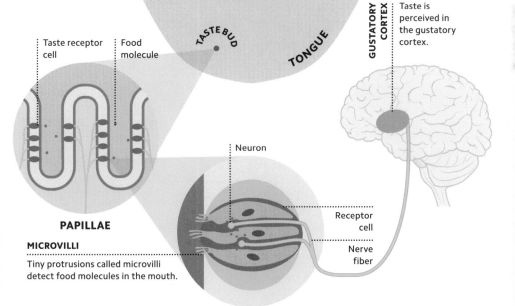

TASTE BUD

TONGUE

GUSTATORY CORTEX
Taste is perceived in the gustatory cortex.

Taste receptor cell

Food molecule

Neuron

Receptor cell

Nerve fiber

PAPILLAE

MICROVILLI
Tiny protrusions called microvilli detect food molecules in the mouth.

TASTE RECEPTOR CELL

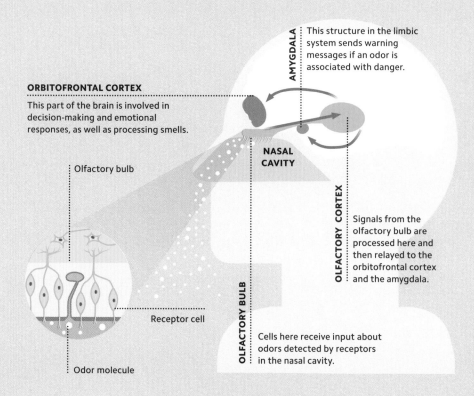

AMYGDALA

This structure in the limbic system sends warning messages if an odor is associated with danger.

ORBITOFRONTAL CORTEX

This part of the brain is involved in decision-making and emotional responses, as well as processing smells.

Olfactory bulb

NASAL CAVITY

OLFACTORY CORTEX

Signals from the olfactory bulb are processed here and then relayed to the orbitofrontal cortex and the amygdala.

Receptor cell

OLFACTORY BULB

Cells here receive input about odors detected by receptors in the nasal cavity.

Odor molecule

ON THE SCENT

The sense of smell depends on roughly 400 different receptor proteins that can be activated by airborne chemicals inhaled into the nose. In the nose, a sheet of 10–20 million olfactory neurons lines our nasal cavities. Each neuron contains just 1 of the 400 smell receptors. The perception of an odor depends on the subset of receptors the odor activates, and therefore the neurons it causes to fire. Olfactory signals activate brain regions that function in emotion and memory, meaning that smells often evoke feelings of pleasure or disgust and become entwined with our memories.

BLENDED SENSES

Synesthesia is when a person experiences two or more sensory perceptions in response to something that is usually experienced as just one. Any combination of senses is possible: colors may be tasted, a brush on the skin may smell, a numeral might be a squeak. The links between perceptions are fixed for life. Brain imaging shows that synesthesia occurs in the brain itself—a person who "tastes" words, for instance, has activity in their gustatory cortex as well as the auditory areas when they hear a word. It is thought to occur in people who, unusually, have neural pathways between different sensory brain areas.

Seeing music

A common type of synesthesia is seeing sounds as colors. A musical note—middle C, say—can evoke the sensation of yellow, or music may be perceived as moving shapes or even smells.

MOODS
EMOTI

AND
ONS

Emotions are generated in the limbic system, the parts of the brain that lie beneath the cortex and above the brainstem. One tiny part of it, the amygdala, constantly receives information about the environment and, if it detects anything that might call for a response—such as a threat—it sends urgent messages to other brain areas. These areas prepare the rest of the body to respond. Moods are emotional states that are less intense but more prolonged than emotions. They color the way we see the world and alter our behavior. Intense moods are commonly regarded as illnesses. Examples include depression (if the state is very sad) or mania (if ecstatic).

HOW DO YOU FEEL?

Human emotions are primarily bodily responses to external events or stimuli. A stimulus will cause the brain to trigger the release of hormones, such as testosterone, adrenaline, and serotonin. These drive changes in our physical reactions—such as by raising heart rate or blood flow—and help prepare us physically for a response. Many emotions are unconscious (see p.47).

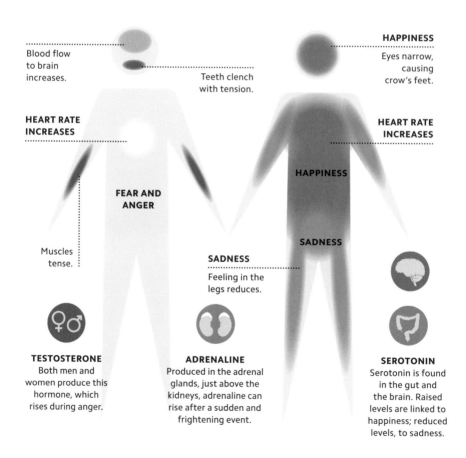

HAPPINESS

Blood flow to brain increases.

Teeth clench with tension.

Eyes narrow, causing crow's feet.

HEART RATE INCREASES

HEART RATE INCREASES

HAPPINESS

FEAR AND ANGER

Muscles tense.

SADNESS

SADNESS
Feeling in the legs reduces.

TESTOSTERONE
Both men and women produce this hormone, which rises during anger.

ADRENALINE
Produced in the adrenal glands, just above the kidneys, adrenaline can rise after a sudden and frightening event.

SEROTONIN
Serotonin is found in the gut and the brain. Raised levels are linked to happiness; reduced levels, to sadness.

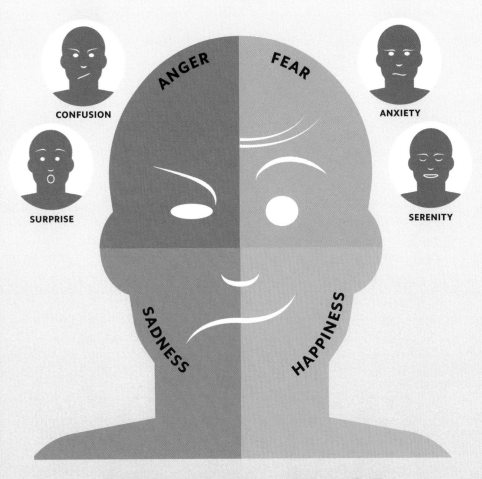

EMOTIONAL RANGE

Scientists recognize four core physiological responses that translate into conscious emotions: anger, fear, happiness, and sadness. These combine to create emotional range. Common emotions such as fear, joy, surprise, sadness, anger, and disgust each link to a specific facial expression. Feelings such as guilt, jealousy, frustration, and nostalgia are mixtures of basic emotions: for example, guilt merges fear (of punishment) with anger and possibly disgust (with ourselves).

Moods are mental states that mirror emotions (see p.42) but that persist, often unconsciously, and are less obviously linked to external triggers. We may not be aware of being in a mood, but the state has subtle effects on our perceptions, thoughts, and behavior. Some people are naturally disposed to moodiness, but most people's brains self-regulate and the person eventually returns to their normal state of mind.

(see p.42)

IN A MOOD

HIGHS AND LOWS
Almost everyone goes up and down in mood. Ecstatic moods are often followed by deep lows.

Part of growing up

Mood swings, where moods change suddenly and unpredictably, are common and normal during adolescence. In adults, regular fluctuations between very high mood (mania) and very low mood (depression) may point to a more serious condition, such as bipolar disorder.

"[Bipolar disorder] is a challenge, but it can set you up to be able to do almost anything else in your life."
Carrie Fisher

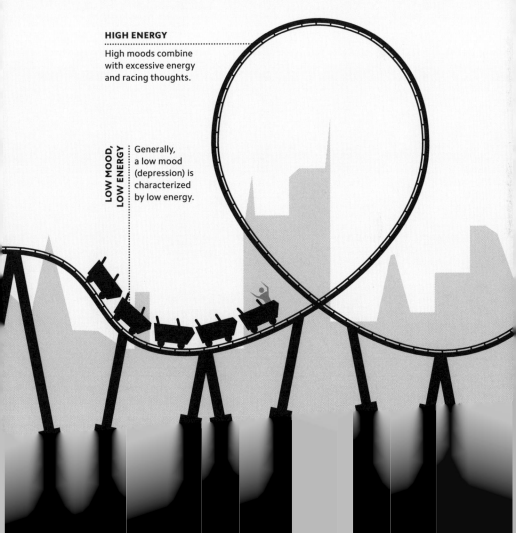

HIGH ENERGY
High moods combine with excessive energy and racing thoughts.

LOW MOOD, LOW ENERGY
Generally, a low mood (depression) is characterized by low energy.

GENERATING EMOTIONS

Emotions are triggered by neural activity in the amygdala—a small almond-shaped organ in the limbic system (see p.16). The amygdala is excited by environmental triggers, such as threats and potential rewards. It also responds to one's thoughts—we can make ourselves sad just by recalling a tragic event. Some studies suggest that negative emotions such as fear, sadness, and anger are generated mainly by the right hemisphere amygdala and positive ones by the left.

AWARENESS

Anything in the environment that demands some sort of response may generate emotion.

Visual cortex

Amygdala

THREAT DETECTED

The amygdala responds by preparing the body for action.

TAKING ACTION

We usually act before we consciously register an emotion or threat—a potentially life-saving reflex.

ANGER

HIDDEN FEELINGS

The amygdala responds to a wide range of stimuli, including others' facial expressions and every unusual sight, sound, and smell. The emotions generated by these things constantly adjust our body so that we are primed to respond appropriately, but most of the time, we are unconscious of them. We "feel" an emotion only when amygdala activity is very intense or prolonged.

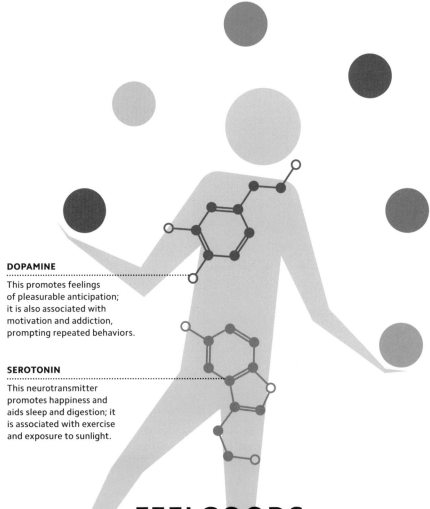

DOPAMINE
....
This promotes feelings
of pleasurable anticipation;
it is also associated with
motivation and addiction,
prompting repeated behaviors.

SEROTONIN
....
This neurotransmitter
promotes happiness and
aids sleep and digestion; it
is associated with exercise
and exposure to sunlight.

FEELGOODS

The amygdala (a brain structure associated with emotional response)
registers external stimuli and works with other brain areas to decide
whether they are likely to be harmful or beneficial. If it calculates that
a stimulus is good—for example, the aroma of food—the amygdala
triggers production of dopamine, a chemical neurotransmitter linked
to feelings of pleasurable anticipation. Also involved in this response is
the release of serotonin, a neurotransmitter that creates feelings of
satisfaction and is produced when an anticipated goal is achieved.

FEELBADS

Negative emotions, such as fear or anger, are produced when we experience or anticipate something harmful. The brain prompts bodily changes to enable fight, flight, or (if effective action isn't possible) freezing—known as the fight-or-fight reflex. These responses involve production of the neurotransmitter noradrenaline (or norepinephrine) as well as the hormone adrenaline (or epinephrine), which work in tandem to help the body and brain react quickly to a threat. However, afterward, we may be left feeling low on energy and irritable.

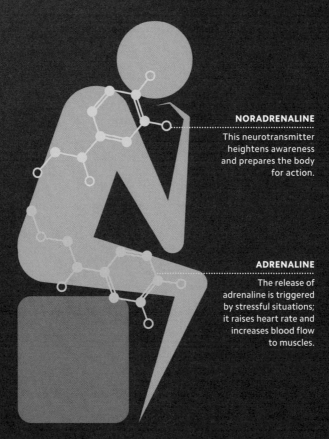

NORADRENALINE

This neurotransmitter heightens awareness and prepares the body for action.

ADRENALINE

The release of adrenaline is triggered by stressful situations; it raises heart rate and increases blood flow to muscles.

ARTISTIC COMMUNICATION

There is no brain area dedicated to aesthetic appreciation. Images on canvas that represent a real object trigger similar neural activity to that which is triggered by looking at the object itself. Visual art is not limited to representations of the world, however. Art may distort natural scenes or depict sights that do not exist outside. The brain activity triggered by artwork involves a circuit known as the default mode network (DMN), which is associated with rumination and imagination.

RESTING STATE

If the brain is not active with a task—for example, when we observe art—it switches to a resting state, activating the DMN and enabling reflection.

> "This world is
> but a canvas to
> our imagination."
> Henry David Thoreau

Wandering mind

A visual representation of an event or a landscape may at first provoke in the viewer a similar brain response to that triggered by the thing itself. However, the brain quickly overrides the initial response and may enter a state that is similar to daydreaming.

DAYDREAMING

The DMN, which is active during daydreaming, triggers regions of the brain associated with thinking of others, self-reflection, and the past and future, all of which form the basis of most daydreams.

MUSIC AND THE BRAIN

Music, like all sounds, is registered in the brain's auditory cortex. It produces responses in both brain hemispheres, though for most people the right hemisphere is activated more than the left. Different aspects of music excite different parts of the auditory cortex. The lower area, where input from the ear first impacts, responds to the frequencies of sound waves. The area above this registers harmony and rhythm, while in the uppermost part, music generates activity linked to memory and meaning. The "Mozart effect" is the alleged intelligence-boosting effect of music by the composer Mozart upon young brains. However, recent research suggests that, while listening to music in childhood helps develop cognitive skills, this is not confined to the works of Mozart.

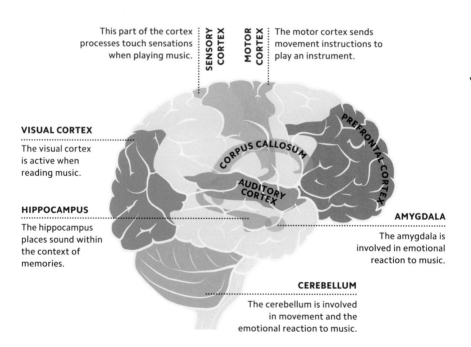

SENSORY CORTEX
This part of the cortex processes touch sensations when playing music.

MOTOR CORTEX
The motor cortex sends movement instructions to play an instrument.

VISUAL CORTEX
The visual cortex is active when reading music.

PREFRONTAL CORTEX

CORPUS CALLOSUM

AUDITORY CORTEX

HIPPOCAMPUS
The hippocampus places sound within the context of memories.

AMYGDALA
The amygdala is involved in emotional reaction to music.

CEREBELLUM
The cerebellum is involved in movement and the emotional reaction to music.

Musical multitasking
Listening to music uses multiple areas of the brain, and even more are used in musical performance, such as playing an instrument or singing. Lyric recall involves the language centers, while reading music activates the visual cortex. Emotional responses are triggered by the amygdala.

INDIVID

UALITY

No two brains are identical. Even identical twins have different brains by the time they are born. The gross anatomy of the brain is similar in neurotypical people, but the neural pathways that determine behavior differ because unique genetic inheritance and experiences determine how neurons connect. The physical differences that distinguish one person's brain from another are usually too small to be seen, but even minute physiological differences cause individual brains to function differently and therefore produce varying behavior. Functional brain changes are also associated with many types of mental disorders and with altered states of mind.

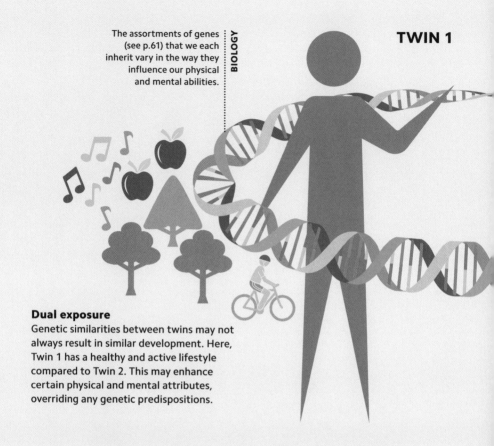

The assortments of genes (see p.61) that we each inherit vary in the way they influence our physical and mental abilities.

BIOLOGY

TWIN 1

Dual exposure
Genetic similarities between twins may not always result in similar development. Here, Twin 1 has a healthy and active lifestyle compared to Twin 2. This may enhance certain physical and mental attributes, overriding any genetic predispositions.

NATURE VS. NURTURE

We are all individuals. But what are these personal differences due to? In the modern version of the age-old nature versus nurture debate, "nature" means the characteristics we inherit in our DNA, while "nurture" refers to our lifetime experiences. For some characteristics or dispositions, how much nature or nurture contributes has been worked out from studies of identical and nonidentical twins. For example, a person's happiness is estimated to be about 50 percent due to genetic factors and about 50 percent due to life's ups and downs.

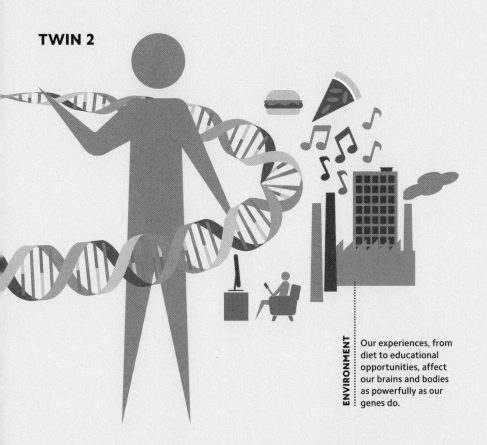

TWIN 2

ENVIRONMENT Our experiences, from diet to educational opportunities, affect our brains and bodies as powerfully as our genes do.

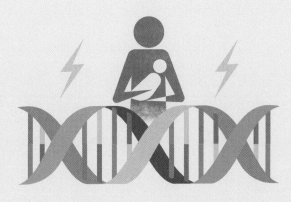

Epigenetic changes
Some experiences change our DNA chemically, altering the way in which genes are "expressed" (switched on or off). These "epigenetic" changes cause the experiences to become physically encoded in us.

THE CHANGING BRAIN

Babies are born with immature brains due to the limit on head size for natural birth to be possible at all. Some innate behaviors, such as the rooting (nipple-seeking) reflex, help them thrive from birth. In childhood, huge numbers of new connections between neurons are made, well-used connections are strengthened, and less-used ones are lost. Some skills, such as learning the sounds of a language like a native speaker, must take place within specific periods of childhood to happen at all. In the teenage and early adult years, the brain is still developing. The prefrontal cortex (see frontal lobe, pp.14–15) matures, enabling higher abilities such as reasoning, planning, and self-control to function.

Birth	Childhood	Adolescence
The brain starts to develop when an embryo is 3 weeks old. At birth, it has an average of 100 million neurons (as many as an adult's).	Brain volume doubles in the first year after birth. A dense network of fibers forms between neurons; this is "pruned" in response to experience.	During adolescence, different brain areas connect and mature. The areas mature at different rates, with the prefrontal cortex doing so last.

Adult

The adult brain is mature, but it continues to change in response to experience. As we get older, our special skills and wisdom continue to improve into our 60s or later.

Aging

"Fluid intelligence"—the ability to learn new skills—reduces with age. So as we age, learning becomes harder, even though many mental abilities can continue.

Size

On average, the male brain is larger than the female brain by around 11 percent. However, the individual brain sizes of different people can still vary, regardless of whether they are male or female.

Connections

Some studies have found more neural connections between hemispheres in the female brain compared to the male. Whether such wiring distinctions are due to genes or to cultural influences remains uncertain.

MALE   FEMALE

MORE ALIKE THAN DIFFERENT

While countless studies have tried to establish whether male and female brains are "made" differently, there are only a few clear findings about the role of genes in brain anatomy. One such finding is that male brains are "naturally" bigger than female brains by more than 10 percent. Studies also show differences in the way that our brains are wired—a female brain has more neural connections between its two hemispheres than a male brain, where the connections tend to be more front to back. This is widely thought to explain why—according to many studies— women are able to access a wider range of information than men when they think about something but are less likely to focus narrowly on it.

GENETIC BLUEPRINT

Our genes determine the broad architecture of our brains, as well as many of the individual quirks that make us different from everyone else. Major neural pathways, such as the one that carries electrical signals from the eye to the visual cortex (see p.32), develop according to the basic human genetic program and are similar in everyone. Genes may also carry instructions for more variable neural features—for example, an individual might inherit the neurochemical "recipe" that produces a tendency to be fearful or reckless. Most cognitive skills are affected both by genes and by the environment in which a person is brought up. Intelligence, for example, seems to depend equally on genetic inheritance and education (see pp.56–57).

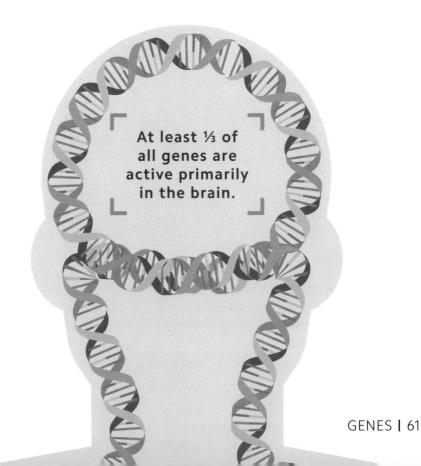

At least ⅓ of all genes are active primarily in the brain.

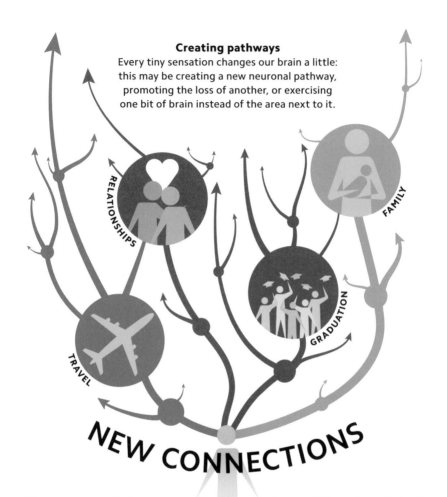

Creating pathways
Every tiny sensation changes our brain a little:
this may be creating a new neuronal pathway,
promoting the loss of another, or exercising
one bit of brain instead of the area next to it.

RELATIONSHIPS

FAMILY

TRAVEL

GRADUATION

NEW CONNECTIONS

The human brain is very sensitive to experience. Although we inherit much of our personality and skills, our environment determines what we are like as much as our genes. The "molding" by experience starts before birth—identical twins are born with distinct brains because each will have occupied a separate position in the womb and therefore will have received minutely different sensations as their mother moved around. Young brains are particularly "plastic" (see pp.24–25), so a child's early experiences impact heavily on their brain. Our brains never stop changing from experience, and people can learn new things at any age.

EVOLVING CHARACTERISTICS

Psychologists define personality traits as our habits of behavior (as well as the thoughts and feelings underlying them) that remain stable over time; these are in contrast to moods or other shorter-term changes. Our genetics account for about half of the variation in personality between individuals. Recent research suggests that some personality traits—for example, emotional stability—are more genetically based than other traits, such as introversion or extroversion. It is now also recognized that our personality is able to change during our lifetime. For example, humans typically become more emotionally stable and conscientious with age. We may even change certain aspects of our personality deliberately. So while personality traits may appear relatively stable, they are not fixed.

FAMILY AND CHILDHOOD FACTORS

EXPERIENTIAL FACTORS

EVOLVING PERSONALITY

GENETIC FACTORS

A growing personality
Our personality develops with age, affected by genetic factors as well as those linked to family and upbringing. Also highly influential are our individual experiences as we go through life.

WHO ARE YOU?

One way of assessing personality is to look at how much an individual has of specific, predefined personality factors. Various systems have been formulated to try to assess and characterize the personality of individuals. Some of the best-known personality tests, such as Myers–Briggs—which cites 16 personality types—are often used in workplaces, and while they are not scientifically validated, they can help people develop better ways of working together.

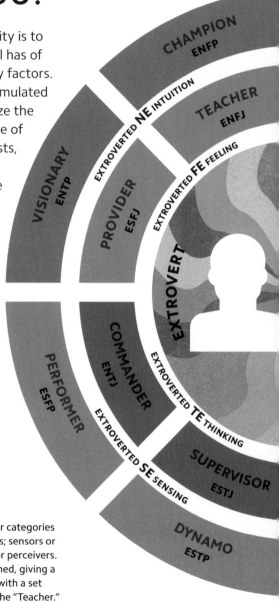

KEY

Extroverts (E)	◄──►	Introverts (I)
Sensors (S)	◄──►	Intuitives (N)
Thinkers (T)	◄──►	Feelers (F)
Judgers (J)	◄──►	Perceivers (P)

CHAMPION
ENFP

TEACHER
ENFJ

EXTROVERTED **NE** INTUITION

EXTROVERTED **FE** FEELING

VISIONARY
ENTP

PROVIDER
ESFJ

EXTROVERT

PERFORMER
ESFP

COMMANDER
ENTJ

EXTROVERTED **TE** THINKING

EXTROVERTED **SE** SENSING

SUPERVISOR
ESTJ

DYNAMO
ESTP

Myers–Briggs model
Participants are rated according to four categories of paired traits: extroverts or introverts; sensors or intuitives; thinkers or feelers; judgers or perceivers. One letter from each category is assigned, giving a four-letter score. This will correspond with a set personality type; for example, ENFJ is the "Teacher."

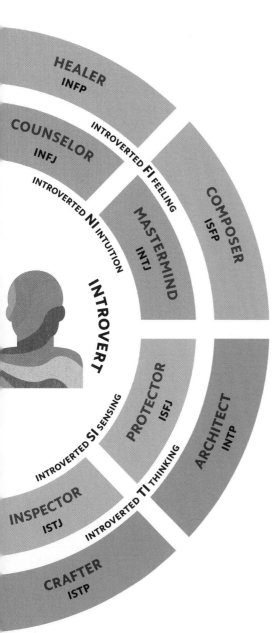

HEALER
INFP

COUNSELOR
INFJ

INTROVERTED **Fi** FEELING

INTROVERTED **Ni** INTUITION

MASTERMIND
INTJ

COMPOSER
ISFP

INTROVERT

PROTECTOR
ISFJ

ARCHITECT
INTP

INTROVERTED **Si** SENSING

INTROVERTED **Ti** THINKING

INSPECTOR
ISTJ

CRAFTER
ISTP

"There is no such thing as a pure introvert or extrovert."
Carl Jung

Traditional		Creative
OPENNESS		
1 (20)		100
Conventional		Imaginative

Spontaneous		Organized
CONSCIENTIOUSNESS		
1	(63)	100
Impulsive		Persistent

Quiet		Enthusiastic
EXTROVERSION		
1	(48)	100
Reserved		Assertive

Cynical		Empathetic
AGREEABLENESS		
1	(79)	100
Demanding		Polite

Relaxed		Tense
NEUROTICISM		
1 (29)		100
Resilient		Sensitive

The Big Five test

The OCEAN model, or Big Five, proposes that personality is composed of five core traits: openness, conscientiousness, extroversion, agreeableness, and neuroticism. Each trait exists on a sliding scale, and a person may exhibit a trait to any degree on that continuum.

MEMO
AND
LEAR

R Y

N I N G

Memory is a process in which a person's experience is physically encoded in the brain and may be used later to guide behavior. Learning is another word for the recall and use of encoded memories. Very little of our experience is memorized; most of it just passes us by. Events that are sensational, unfamiliar, or accompanied by strong emotion tend to "stick," however, and almost anything can be committed to memory if you attend to it and recall it frequently. Frequent recollection can falsify a memory, though, because every time we consciously remember something, we add, change, or lose some of it.

TYPES OF MEMORY

Memory refers to the process of taking information from the world around us and storing, retaining, and later retrieving this information. Our memories are unique to each of us and shape us as individuals. There are several types of memory, each with a different function.

MEMORY SYSTEMS

Memories can be grouped into short-term memories and long-term memories. Short-term memory includes working memory, the type of memory you use when you want to remember something without writing it down. Long-term memories can last a lifetime and are divided into two groups: declarative, which are conscious, and nondeclarative, which are unconscious.

SHORT-TERM MEMORY

"Whenever you read a book or have a conversation, the experience causes physical changes in your brain."
George Johnson

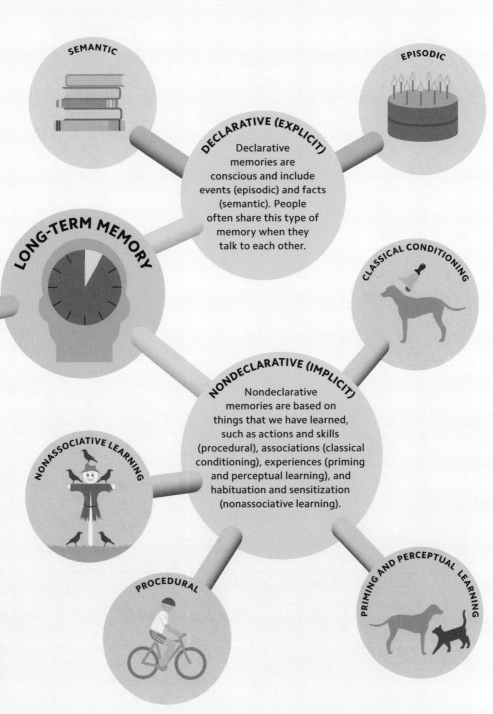

SEMANTIC

EPISODIC

DECLARATIVE (EXPLICIT)
Declarative memories are conscious and include events (episodic) and facts (semantic). People often share this type of memory when they talk to each other.

LONG-TERM MEMORY

CLASSICAL CONDITIONING

NONDECLARATIVE (IMPLICIT)
Nondeclarative memories are based on things that we have learned, such as actions and skills (procedural), associations (classical conditioning), experiences (priming and perceptual learning), and habituation and sensitization (nonassociative learning).

NONASSOCIATIVE LEARNING

PROCEDURAL

PRIMING AND PERCEPTUAL LEARNING

STORING MEMORIES

Memory storage relies on different parts of the brain working together. Explicit memories are stored using the hippocampus, the cortex, and the amygdala. These memories are temporarily stored in the hippocampus and later transferred to the cortex for long-term storage. The amygdala attaches emotions to our memories. It is thought that much of our memory consolidation happens during sleep.

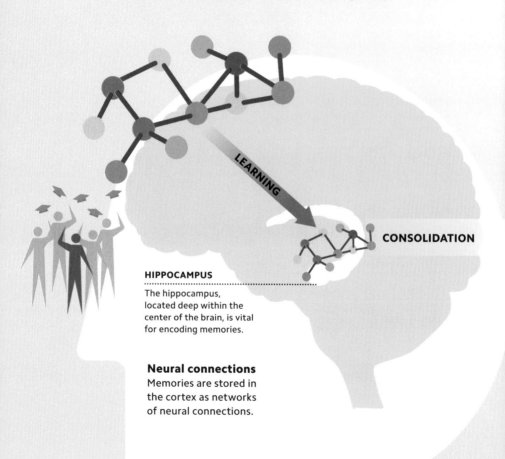

LEARNING

CONSOLIDATION

HIPPOCAMPUS

The hippocampus, located deep within the center of the brain, is vital for encoding memories.

Neural connections

Memories are stored in the cortex as networks of neural connections.

RECALLING MEMORIES

Memory recall is the final step in processing a memory. Recall allows us to remember the information or events that have previously been encoded and stored in the brain. To recall a memory, the brain replays the event that you are trying to remember, and the stronger the pathway that has been encoded, the faster the memory can be recalled. The shorter the time between creating a memory and recalling the memory, the more likely the memory will be recalled successfully.

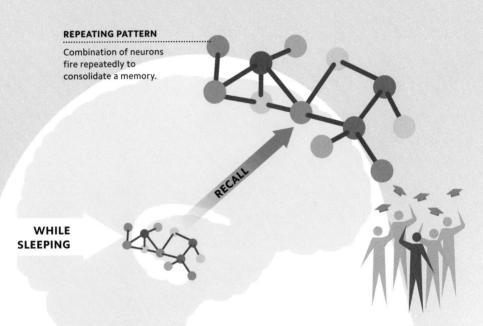

REPEATING PATTERN
Combination of neurons fire repeatedly to consolidate a memory.

RECALL

WHILE SLEEPING

Strengthened connections
The more often a memory is recalled, the stronger the neural connections that encode that memory become. However, each recall can alter a memory (see p.75).

IDENTIFYING THE FAMILIAR

Recognition gives that "aha!" feeling of familiarity. Although it may seem instantaneous, it is a multistep process involving many brain areas. When you see something you have seen before, the visual information shunts forward from the visual cortex along nerves in the lower left hemisphere (the fusiform gyrus). This is known as the object recognition pathway. The sight of a face activates a specialized area along the way (the facial recognition area). As it travels, the information is linked to associated memories. A face is linked to a name, for instance, and an object to its function.

A face in the crowd
Recognition may be partial. A face in a crowd might "ring a bell," but you may not be able to "place" it.

MONTHS

Storage
Memories are encoded in the hippocampus and then stored longer term in the cortex (see p.70).

YEARS

Memory fades
If a memory is not recalled, it will begin to fade, becoming less detailed and vibrant.

DECADES

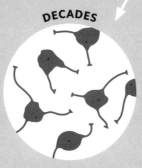

Losing a memory
Eventually, memories can become inaccessible. New research suggests that forgetting may be a functional feature of the brain.

WHY WE FORGET

Memories are made when neurons get linked together because they are repeatedly activated simultaneously. The links may be weak (forming a short-term memory) or strong (long-term). Long-term memories may last a lifetime, but if they are not kept "warm" by recall, even the strongest neural links may decay over time, and when they have gone, the memory has gone, too. Forgetting something is a natural process of the human brain, and we all experience different stages of natural memory decay.

IMPROVING YOUR MEMORY

There are various ways to help our memory. When we are trying to learn something, it helps to take regular breaks that involve some physical exercise. Repeating things over and over (rehearsal) helps consolidate them. Mnemonics are ways of linking new information to something familiar or easy to remember—such as a sentence, a phrase, or even a location you know well—so that when you recall the familiar things, you also bring out the new information. Memories are consolidated during sleep, particularly deep sleep.

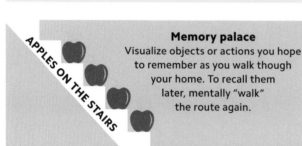

APPLES ON THE STAIRS

Memory palace
Visualize objects or actions you hope to remember as you walk though your home. To recall them later, mentally "walk" the route again.

PINEAPPLE IN THE DINING ROOM

FISH IN THE BASEMENT

FALSE MEMORIES

False memories are memories of things that did not actually happen but seem to have. They can be entirely fabricated or contain elements that are true or that have been distorted from the truth. Everyone has false memories because our memories are malleable and can be influenced. Every time a memory is recalled, it can be altered by adding elements from the present moment. How we are asked about a memory can also influence how we remember it. For example, if we are asked a leading question such as, "Do you remember the red t-shirt she wore?," we are likely to imagine that we did, even if there was never a red t-shirt.

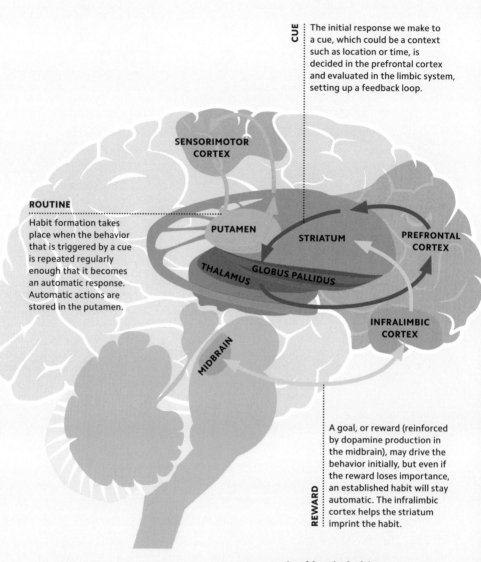

CUE The initial response we make to a cue, which could be a context such as location or time, is decided in the prefrontal cortex and evaluated in the limbic system, setting up a feedback loop.

SENSORIMOTOR CORTEX

ROUTINE
Habit formation takes place when the behavior that is triggered by a cue is repeated regularly enough that it becomes an automatic response. Automatic actions are stored in the putamen.

PUTAMEN

STRIATUM

PREFRONTAL CORTEX

THALAMUS

GLOBUS PALLIDUS

INFRALIMBIC CORTEX

MIDBRAIN

REWARD A goal, or reward (reinforced by dopamine production in the midbrain), may drive the behavior initially, but even if the reward loses importance, an established habit will stay automatic. The infralimbic cortex helps the striatum imprint the habit.

Locking in habits
Many brain areas regulate and maintain goal-directed and habitual behaviors. Central to this is the striatum, which coordinates three pathway loops that assess a new goal, enable a linked behavior, and imprint it as habit with repetition.

MEMORIES THAT AREN'T FORGOTTEN

Habits are ingrained, automatic behaviors that do not require conscious thought. They usually involve a trigger (or cue) that is linked to a particular behavior. This behavior triggers the brain's reward system: a complex network of pathways that alter the brain's chemistry. As a result, the behavior is then associated with feeling better. The association is strengthened with each repetition of the behavior, making it difficult to stop. Unhealthy habit formation is a defining trait of many psychiatric disorders, such as obsessive-compulsive disorder (OCD) and addiction.

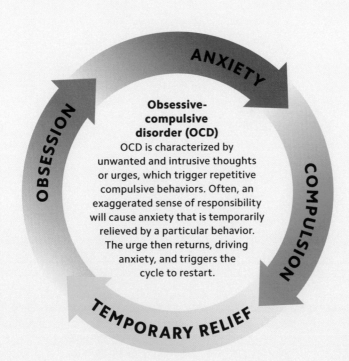

ANXIETY

COMPULSION

TEMPORARY RELIEF

OBSESSION

Obsessive-compulsive disorder (OCD)
OCD is characterized by unwanted and intrusive thoughts or urges, which trigger repetitive compulsive behaviors. Often, an exaggerated sense of responsibility will cause anxiety that is temporarily relieved by a particular behavior. The urge then returns, driving anxiety, and triggers the cycle to restart.

COMMUN

ICATION

Like many other animal species, humans communicate by facial expressions, gestures, and actions. Uniquely, we also use language. The difference between verbal language and other means of communicating is that words allow us to think about and communicate abstract ideas instead of limiting us to the concrete world. Through verbal language, we share our thoughts, intentions, memories of the past, and visions of the future. Language is underpinned by brain structures that are unique to humans. In most people, these structures are found in the left hemisphere, while gestures, rhythm, and tonal expression engage the right side.

3 MONTHS +

PLOO!

COO!

Hardwired bias
When babies are born, they prefer animal vocalizations—made by both humans and monkeys—to other noises. Then, within three months, they fine-tune their preference to the sound of the human voice, even if it is speaking nonsense words.

THE LANGUAGE BLUEPRINT

Species communicate with each other in a variety of ways—many through vocalized sounds or physical signs. However, human brains are fully adapted for language processing at a much higher level than the rest of the animal kingdom. Only humans use words—which can refer to things that are not present—and syntax: language grammatically structured into phrases and sentences. This innate ability allows us to share ideas; make plans; and refer to the past, present, or future, all of which is invaluable to human advancement.

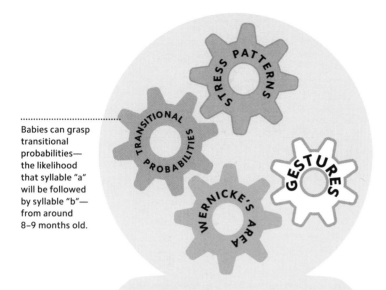

Babies can grasp transitional probabilities—the likelihood that syllable "a" will be followed by syllable "b"—from around 8–9 months old.

TAKING IT ALL IN

Human brains have evolved a language "instinct," so children normally develop it naturally. Usually, parents or carers also help babies understand words by gesturing to or miming their meaning. The stress patterns of phrases also help communicate their meaning to babies. Once the meaning of a word is learned, the brain retrieves it from memory when the sound of it (or sight of it, if it is sign language) is encountered again. This process involves Wernicke's area (see pp.82–83). At birth, babies' brains are primed for any language. However, if they only hear one tongue spoken, the auditory neurons that respond to foreign tones die off. Once this has happened (from 3 years of age), it is very difficult to learn to speak another language without an accent.

VERBAL AND NONVERBAL

Language is a built-in skill that develops naturally in children, provided they hear people speaking around them. The main language areas of the brain lie in the left hemisphere in most people. They are Wernicke's area, which converts the sound of words into their meaning, and Broca's area, which controls the physical movements needed to speak. Wernicke's area is located close to the junction of the temporal and the parietal cortex—next to the part that registers sound. Damage to a person's Wernicke's area prevents them from understanding words, but it doesn't inhibit their speech. As a result, the individual might speak perfectly clear but nonsensical words. Broca's area is in the frontal lobe, near the motor area controlling the mouth and tongue. Damage to Broca's area stops the person from speaking fluently but does not affect their understanding.

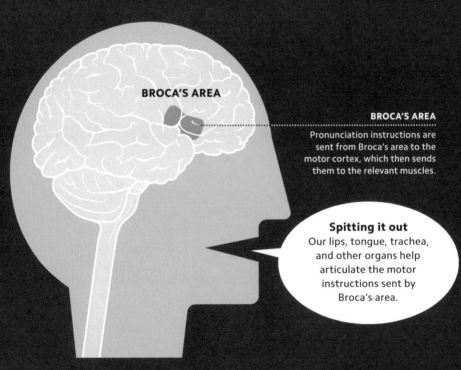

BROCA'S AREA

BROCA'S AREA

Pronunciation instructions are sent from Broca's area to the motor cortex, which then sends them to the relevant muscles.

Spitting it out
Our lips, tongue, trachea, and other organs help articulate the motor instructions sent by Broca's area.

Speaking with your body

Sign language is a visual way of communicating. Different sign languages have different grammars and structures, even if their spoken versions are similar, such as English Sign Language and American Sign Language (pictured). This is because sign language is not a literal translation of speech. However, just like in spoken conversations, Broca's area and Wernicke's area activate when an individual is signing. Conversely, signing naturally engages visual areas in the brain instead of auditory areas.

HELLO

PLEASE

Understanding

Recognizing words and giving them meaning occurs in Wernicke's area.

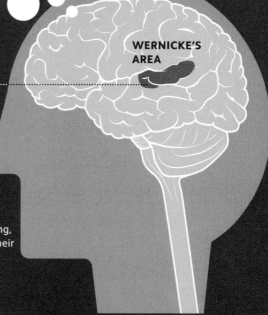

WERNICKE'S AREA

WERNICKE'S AREA

When we listen to speech, cochlear receptors pass audio signals on to the auditory cortex. This sends auditory information to Wernicke's area.

Brain meets brain

When a person is speaking, signing, or gesturing, different parts of their brain are active compared to the brain of the person on the other end of the interaction who is hearing or observing them.

Contracted pupil
Our pupils contract in strong light or when focusing on a near object. They can also contract when we feel relaxed, angry, or disgusted.

SPEAKING WITHOUT WORDS

Body language refers to the nonverbal signals that we use to communicate. These signals make up a large part of daily communication even if we are not always aware of them. Reading someone's posture, gestures, or movements might seem simple. However, processing body language is highly complex and involves specific regions in the brain, including the amygdala (where emotions are processed), visual areas, and areas linked to planning and executing actions. While the recognition of facial expressions (see p.43) is well studied, the processing of body language has received less attention until recently. Changing our body language can also influence our thoughts and feelings.

Dilated pupil

Our pupils dilate in low light. This can also occur when we are surprised, putting mental effort into something, or feeling attracted to someone.

Holding your arms up and your chest out for just two minutes can boost self-confidence and assertiveness and reduce stress.

THE WRITTEN WORD

Unlike speech, the ability to read and write (literacy) is not "wired in" and must be taught. Becoming literate involves creating new neural connections between brain areas dedicated to language (in the left hemisphere in most people) and others involved in decoding shapes and sounds. Reading words involves decoding symbols into spoken words and then processing them just as we would if they had been spoken. This is why unpracticed readers mouth words as they read. Writing requires doing the same thing backward: first we "hear" the words in our head, then our brain translates them into learned symbols, before finally instructing our muscles to make the movements that put them on the page.

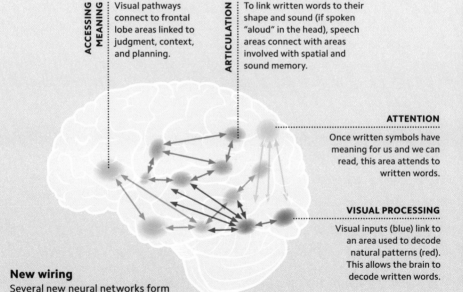

ACCESSING MEANING
Visual pathways connect to frontal lobe areas linked to judgment, context, and planning.

ARTICULATION
To link written words to their shape and sound (if spoken "aloud" in the head), speech areas connect with areas involved with spatial and sound memory.

ATTENTION
Once written symbols have meaning for us and we can read, this area attends to written words.

VISUAL PROCESSING
Visual inputs (blue) link to an area used to decode natural patterns (red). This allows the brain to decode written words.

New wiring
Several new neural networks form when we learn to read or write. They link many areas, including those associated with visuals, speech, sound, attention, and planning.

Literate

When a literate person reads a sentence, their brain shows increased activity in language areas. There is also increased activity in the auditory part of those language areas, both when reading a sentence and when hearing a sentence spoken. This indicates that, for people who are literate, words are "heard" internally.

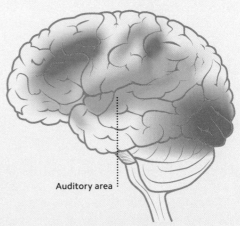

Auditory area

Newly literate

Newly literate brains are activated in the same areas as literate brains. When someone learns to read and write, their brain gets rewired. This can happen at any age and involves areas associated with reading and beyond. For example, when shown a sentence, literate and newly literate readers activate areas that are involved in controlling attention.

Illiterate

An illiterate brain does not react strongly to written words because it is unable to extract meaning from them. The brain's auditory cortex is not stimulated, as it cannot turn the marks on the page into imagined speech. The more tissue there is in a brain, the longer it takes for disease to destroy it. Because learning to read and write creates denser connections in the brain, literacy helps protect from dementia.

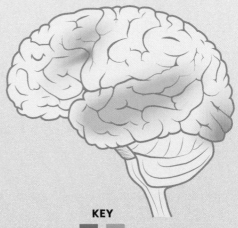

KEY

Written sentences ■ ■ Spoken sentences

THE

BRAIN

ACTIO

A N D

N

The brain's main survival strategy is prediction: it takes in information from the environment and uses it to work out the best move to make in each situation. Action plans are created in the front of the brain, then sent back to the motor cortex. From here, instructions are sent to muscles via the long nerves that run down the spine, producing movement. Creating coordinated, effective action involves many areas of the brain. Much of it is done unconsciously; usually, only the end goal of an action is conscious. Opening a door may be a conscious decision, but the steps (such as locating and manipulating the handle) are usually carried out without conscious calculation.

ORGANIZED MIND

Two brain areas contain cortical "maps" of the body: the motor cortex and the somatosensory cortex. Each area forms a strip that lies like a headband over the top of the brain, with the motor cortex in front. They are divided into sections, each one made of the compressed nerve ends that receive or send signals from or to a particular body part. In each, the sections relating to the hand and face are the largest (see p.36) and the trunk section is the smallest.

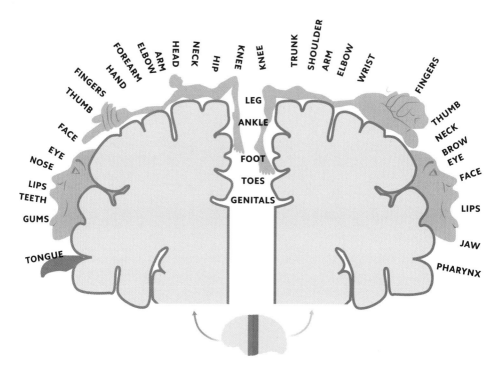

Sensory area
The somatosensory cortex receives information about touch and limb position.

Motor area
The motor cortex generates commands for voluntary muscle movements.

Another hypothesis for this phenomenon suggests that the internal "map" of the body, generated in the brain, still includes the intact limb.

PHANTOM SENSATIONS

Cortical remapping

The cortical area that received messages from a missing body part can be "taken over" by cells that respond to an intact body area. For example, the proximity of the hands and face on the somatosensory cortex "map" means a brush on the face can produce a sensation in a missing hand.

When a limb is amputated or otherwise lost, the neurons that once controlled it and received sensations from it remain. It is common for amputees to live (either briefly or long-term) with a sense that the lost limb is still present—some feel touch or itch sensations, others feel it moving, and some suffer phantom limb pain. Despite the name, phantom limb doesn't only occur in arms and legs and has also been recorded after the loss of other body parts like breasts, eyes, and even teeth.

DISTORTED REFLECTION

In body dysmorphic disorder (BDD), a person thinks their body is flawed, even when it is typical, and they see the imagined flaw in their reflection. Brain imaging studies show that in such people, the brain distorts their perception in the first 200 milliseconds of registering it—well before it is conscious. The visual cortex overresponds to details instead of the overall impression, effectively magnifying the imagined defect.

A 2016 study estimated that nearly 2% of people have a form of BDD.

Bigorexia
Muscle dysmorphia is a particular type of BDD that more commonly affects men. Individuals with the disorder perceive themselves as small and weak when they actually look normal or even very muscular.

CONTROL AREAS

Conscious actions are decided in the brain's frontal lobes, where information from the rest of the brain is brought together and juggled to arrive at the best response. The action plan is formulated in the motor area, which sends the instructions out via long nerves to muscles.

FRONTAL LOBE

Taking in information from the rest of the brain, the frontal lobe determines the appropriate response.

PREMOTOR AREAS

Ideas are turned into movement intentions by this part of the brain before an action plan is created.

MOTOR CORTEX

The supplementary motor cortex (see p.134) passes instructions to this area, which sends signals to the muscles.

CEREBELLUM

The cerebellum monitors and regulates activity, receiving feedback from the body to fine-tune movement.

BRAINSTEM

Signals flow through here to the spinal cord, taking instructions to muscles and bringing back feedback for further movement.

STRIATUM

This group of brain areas helps select movements according to feelings of reward or pleasure and familiarity.

The brain processes many sources of information to guide behavior. Before a ball is even struck, the opponent's posture and paddle swing give anticipatory cues.

To respond optimally, skilled players use stored, proven action sequences. Anticipating the ball's likely trajectory, the premotor and motor cortices prepare to execute those movements.

MAKING A MOVE

Visual information from the eyes travels back to the visual cortex, from which it is passed forward to other brain areas along two main paths: one fast, one slower. The fast pathway travels through the parietal lobes, which calculate the relationship of our body to the outside world. This information passes to the premotor area and motor cortex, producing rapid reactions. The slower pathway travels through the temporal lobe, where it picks up any relevant memories. Information from both pathways combines in the frontal lobe, where a conscious decision about action is made. This may inhibit a movement that has already been started unconsciously and substitute a more considered response.

Based on early data, the motor cortex initiates an action plan, sending signals via the spinal cord to the muscles. These arrive within 100 milliseconds.

Much of this process occurs preconsciously, our conscious minds only catching up after our movements are underway.

As the ball approaches, the brain constantly monitors incoming sensory information. If the trajectory differs from what was anticipated, the motor cortex adapts the action plan.

Rapid action responses

Responding to an event, such as returning a table tennis ball at speed, requires a coordinated response from different brain areas. The receiver must focus attention on the stimulus, combine memory of past action sequences with incoming sensory information, then activate the motor area.

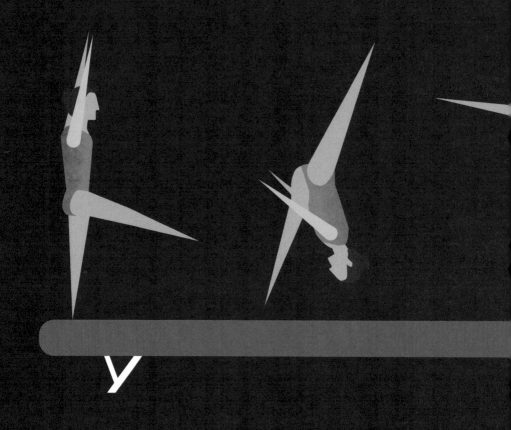

KNOWING YOUR PLACE

Proprioception is the sense of where our limbs and body are in space, and kinesthesia is the awareness of their movement. They are interrelated, largely unconscious senses that allow us to move fluently and to interact easily with objects. They depend on receptor neurons in muscles and joints carrying sensory information to the nervous system about how contracted the muscles are and how extended the joints are. Though rare, some people have no proprioceptive system. They instead track their limb positions visually, moving well when they can see but completely uncoordinated when they cannot.

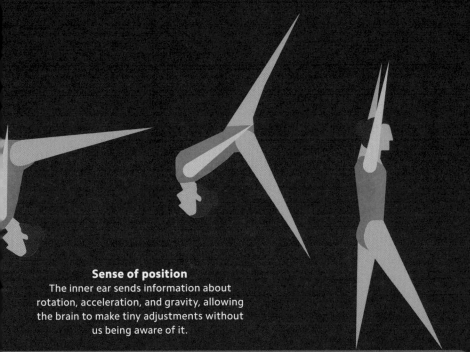

Sense of position
The inner ear sends information about rotation, acceleration, and gravity, allowing the brain to make tiny adjustments without us being aware of it.

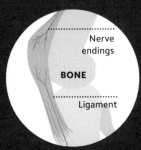

Nerve endings

BONE

Ligament

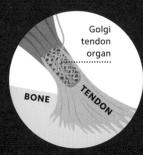

Golgi tendon organ

BONE

TENDON

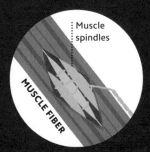

Muscle spindles

MUSCLE FIBER

Joint receptors
Nerve endings within our joints detect the joints' position. They prevent damage by guarding against overextension of the joint.

Tendon receptors
Golgi tendon organs are found within the tendons at the ends of muscles. They monitor muscle tension to ensure they are not overstretched.

Muscle receptors
Muscles have sensors called spindle fibers within them. The spindles send information to the brain about the positions of the muscles.

Broad brushstrokes
Cortical motor commands only outline broader movements, as is evidenced by damage to the cerebellum causing jerky, uncertain movements.

COORDINATION AND FINE-TUNING

While the frontal lobe controls broad, or gross, movement, the cerebellum (also called the "little brain") is responsible for controlling our balance and fine-tuning our movements. In constant communication with the body and the cortex, the cerebellum compares ongoing commands to feedback it receives from the body. It then adjusts the commands to produce smooth, fluid movements. The cerebellum sits at the back of the brain, and while it accounts for only about 10 percent of the brain's volume, it contains over 50 percent of the brain's total neurons.

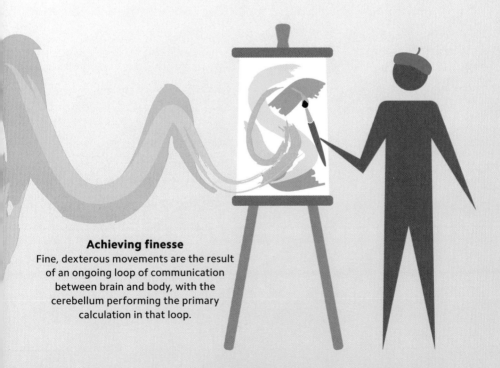

Achieving finesse
Fine, dexterous movements are the result of an ongoing loop of communication between brain and body, with the cerebellum performing the primary calculation in that loop.

THE SOCI

BRAI

A L
N

Humans are social animals, and a significant part of our brain is dedicated to ensuring we get along with each other. Our wide, highly evolved range of social skills make it possible for us to work together on complex, long-term projects and to coexist in dense, interactive communities. Theory of mind—the brain-based ability to know, automatically, something of what is going on in someone else's mind—is a key faculty and underpins the ability we have to empathize with another person. Morality—an idea about what is right and wrong—is also rooted in the brain, even though moral codes differ widely from culture to culture.

> "By six months ... babies are able to recognize a broad variety of faces, including those of another species."
> Oliver Sacks

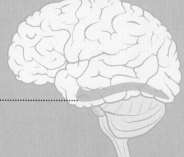

FUSIFORM FACE AREA
This area activates when we look at faces but not when we look at objects.

DO I KNOW YOUR FACE?

Recognizing faces is so important for humans that the brain has an area solely devoted to the task: the fusiform face area within the fusiform gyrus. Brain imaging studies demonstrate that this region becomes active when people look at faces. Damage to or impairment of the right fusiform gyrus is thought to be the cause of prosopagnosia (face blindness)—a neurological disorder characterized by the inability to recognize faces. Some of those affected only struggle to recognize a familiar face; others, to discriminate between unknown faces; and others still, to distinguish a face from an object. Some people with the disorder are even unable to recognize their own face.

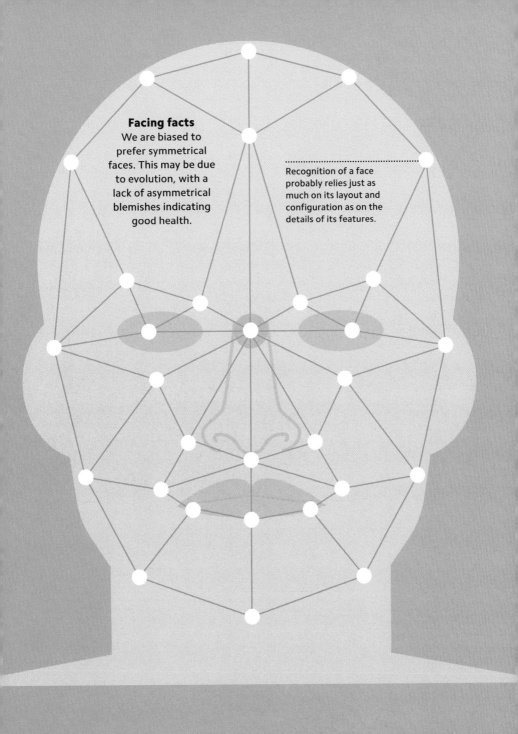

Facing facts
We are biased to prefer symmetrical faces. This may be due to evolution, with a lack of asymmetrical blemishes indicating good health.

Recognition of a face probably relies just as much on its layout and configuration as on the details of its features.

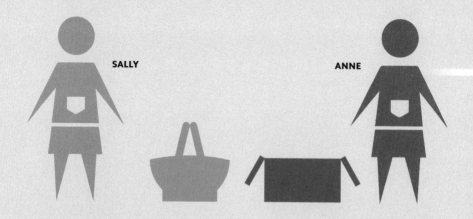

SALLY

ANNE

THE OTHER'S PERSPECTIVE

Theory of mind is our brain's ability to surmise other people's intentions, desires, and beliefs and to take their perspective. It allows us to understand and predict other people's behavior through as little as seeing a facial expression or the look in someone's eyes. This capacity to understand someone else's mind, as well as our own, allows us to enjoy effective human interactions and underpins the ability to empathize with others. Psychopaths (see p.121) can succeed in understanding what others might think and feel but not share their emotions or any concern for them. Autistic people (see pp.126–127), in contrast, may feel empathy but can struggle to recognize someone else's point of view.

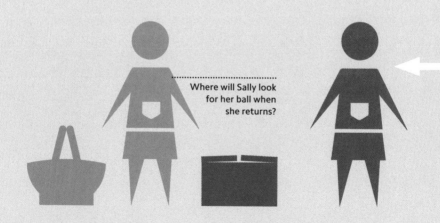

Where will Sally look for her ball when she returns?

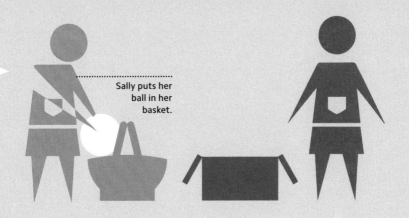

Sally puts her ball in her basket.

The Sally-Anne Test

This psychological test is given to children to assess their theory of mind. For a child to answer the question correctly, they must take Sally's perspective and understand her likely belief that the ball is where she left it—in the basket. This shows theory of mind. However, a child without theory of mind cannot take Sally's perspective and will wrongly assume that Sally knows, as they do, that Anne moved the ball.

Anne moves the ball to her box.

Sally goes away.

READING OTHERS

Reading a person's mind, which relies on theory of mind (see pp.104–105), allows us to understand, explain, predict, and influence the person's behavior. Reading minds is akin to reading text (see pp.86–87): it involves extracting meaning from signs and it activates specific parts of the brain, including the medial prefrontal cortex, the temporal-parietal junction, and the precuneus. Unlike reading text, however, which a child is commonly able to do from 5 years old, basic mind-reading skills do not have to be taught and are usually intuitive. They develop further with help from parents. As carers explain people's thought processes to babies, their mind reading becomes more nuanced.

MONKEY SEE, MONKEY DO

Mirror neurons are brain cells that respond the same way when we observe someone performing an action as when we perform the action ourselves. This internal mirroring is thought to help us learn from others through observation. Mirror neurons respond to body movements, facial expressions, and body language. They were first discovered by accident, when a laboratory monkey whose brain was linked to a device that monitored its activity watched human researchers eating lunch. The monkey's brain activity as it watched matched the activity recorded earlier when the monkey itself reached out and ate food. Mirror neurons are now known to exist in humans and many other species, too.

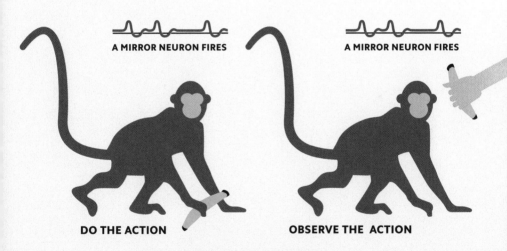

A MIRROR NEURON FIRES

A MIRROR NEURON FIRES

DO THE ACTION

OBSERVE THE ACTION

Not just for humans
Many animals are thought to have mirror neurons. For example, when a monkey witnesses somebody take a banana, its mirror neurons fire as if it is taking the banana.

FEELING WITH

Empathy is the ability to step into another person's shoes and feel "with" them instead of "for" them (which is sympathy). Brain imaging shows that empathy can be so profound that when you see someone you love being physically hurt, the areas in your own brain that produce pain are activated. You literally feel it yourself, although faintly. Empathy is generated by mirror neurons (p.107) and fueled by oxytocin (see opposite).

To the rescue
Seeing someone in pain produces similar brain activation in an empathetic viewer as in the person who is suffering. This motivates the person watching to rescue the other, putting an end to their own discomfort as well as that of the person they help.

When we witness someone else in pain, we may unconsciously mirror their facial expressions and body language.

A DEEP CONNECTION

Forming close connections with others is essential for our well-being. Social bonding is supported by oxytocin, a hormone primarily involved in childbirth and breastfeeding. The same hormone also fuels mother-infant bonding, romantic attachment, altruism, empathy, and sexual arousal. Oxytocin helps us trust and bond with others by reducing the stress response of the prefrontal and limbic cortical circuits (see p.16), which create social anxieties. Oxytocin is also an anti-inflammatory. We are hardwired to be deeply connected with some but divided from others. While bringing us closer to our in-group, oxytocin can make us more aggressive, boastful, and envious of our competitors.

Chemical rewards
Chemicals associated with the reward circuit and with positive emotions—including dopamine, serotonin, and oxytocin (pictured)—are released when we feel close to someone, conditioning us to seek attachments.

$C_{43}H_{66}N_{12}O_{12}S_2$
OXYTOCIN

To follow or not to follow?
Our survival instincts drive our need to belong, as there is safety in numbers. Individuals who challenge the status quo may not be accepted by the group.

ARE WE JUST SHEEP?

The need to belong is fundamental to human identity and influences how we think, feel, and express ourselves. We are more likely to attend to the thoughts and emotions of those in the in-group (the group that we belong to) than to outsiders. We trust and empathize more readily with members of the in-group. Furthermore, often we are even prepared to ignore and override our own views if they conflict with the in-group consensus. While we all have these biases, we are also all capable of overcoming them and breaking from herd behavior. Doing so, however, is often more difficult than not. For example, studies show that when we have some down time, we think about our relationships and our place in the social hierarchy. This occurs when a set of interconnected areas called the default mode network activates. However, when we feel left out or socially isolated, our brains mobilize areas associated with pain.

Engaging in social
behavior triggers
a "bliss" response
through the release
of the neurochemical
oxytocin in the brain.

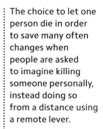

The trolley problem
In this well-known moral dilemma, a person can either pull a lever to divert an out-of-control trolley, choosing to kill one person, or do nothing, saving one person but killing several people. Studies have shown that most people would choose to save the crowd.

The choice to let one person die in order to save many often changes when people are asked to imagine killing someone personally, instead doing so from a distance using a remote lever.

THIS FEELS WRONG

Morality—judging right from wrong—is something that we know instinctively and develop through socialization. While some moral standards vary throughout history and cultures, much of our moral wiring has its basis in our brains and is present from a very early age. Within the first few months of life, babies show a preference toward those acting in a kind and helpful manner and want to see them being rewarded over those being mean. Toddlers make spontaneous efforts to help or console others, even when it may mean they lose out. Moral decisions driven by emotions or empathy (see p.108) result from activity in the orbital and ventromedial parts of the prefrontal cortex.

ALL BY MYSELF

A lack of meaningful social connections can elevate and sustain the body's stress response. When prolonged, this can have mental and physical costs, including high blood pressure, poorer cognition, immune system dysfunction, and higher risk of Alzheimer's disease. Long-term loneliness is comparable to smoking or obesity for its negative impact on our health. Feeling lonely also makes it harder to connect with people. Studies have shown that lonely people are more inclined to pick up on negativity in social interactions. Social contact can also be a form of pain relief. Brain scans have shown that when given mild electric shocks, participants experienced less pain and stress when holding the hand of someone they trusted.

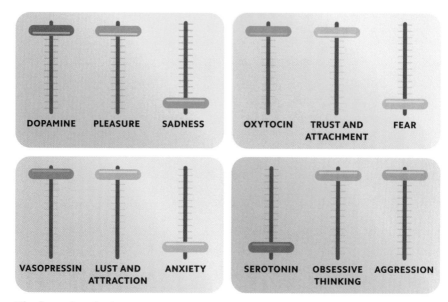

The love chemicals
The level of chemicals in our brain changes when we are in love. These changes heighten some feelings, such as lust and attraction, and lower others, such as fear and anxiety.

LOVE ON THE BRAIN

Falling in love has three main components: lust (sexual desire), attraction, and attachment. Attraction—the feeling of wanting to be with someone—involves reward pathways in the brain and high levels of dopamine. Lust and attraction both lower activity in the prefrontal cortex, reducing our ability to reason. Attachment—the component that holds long-term relationships together—involves the chemicals vasopressin and oxytocin. The latter creates a strong sense of bonding to people we identify with. It also creates a psychological boundary between "us" and "them" (see p.109). While oxytocin makes us feel loving, it takes the other elements of lust and attraction to make us feel romantically in love.

BORN THIS WAY

Whether there is a relationship between brain chemistry and gender and sexuality is a contentious issue. However, there is some evidence that a person's gender identity or sexuality may be tied to prenatal conditioning. In the womb, a fetus is exposed to baths of testosterone that sexualize the brain. The level of testosterone the fetus is exposed to may be a factor in the development of the person's gender or sexuality later in life. In terms of brain structure, some similarities have been observed across a variety of identities—for example, one study found that gay men and heterosexual women had similar brain symmetry—but the significance of these findings is unclear.

A kaleidoscope of possibilities
There is a vast array of gender identities and sexual orientations. They intersect in a variety of different combinations.

NEURODI

VERSITY

Neurodivergent people have brains that process information and learn in a different way to what is considered typical, and they may exhibit atypical behaviors. These behaviors are sometimes due to the way the brain has developed, but some can be seen as typical behaviors that do not match the situation. Schoolchildren who are hyperactive, for instance, may be exploring their environment in a way that is useful at an early age but conflicts with the demands of schooling as they get older. Examples of neurodiversity include autism, attention deficit disorder, difficulty with numbers (dyscalculia), difficulty with reading and writing (dyslexia), and genius.

WE'RE NOT ALL THE SAME

The term neurodiversity is used to acknowledge the wide range of variation in how different people's brains work. Neurodivergent people have brains that function in atypical ways. Examples include attention deficit disorder (ADD), attention deficit with hyperactivity (ADHD), genius (see p.122), and lack of (or excessive) empathy. Other common neurodivergent conditions include dyslexia, dyspraxia—difficulty in coordinating actions and movements—and autism spectrum disorder (see pp.126–127). While some neurodivergent people may need day-to-day support or require adjustments that take into consideration their ways of processing information, for many, neurodiversity means they are particularly good in certain areas and outperform others.

ASD Autistic spectrum disorder (ASD) affects social skills, especially intuitive understanding of other people's beliefs and intentions.

ADD People with attention deficit disorder (ADD) often find it hard to focus, plan, or organize things.

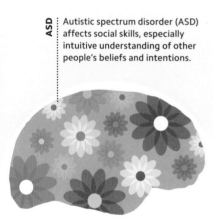

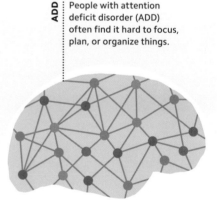

> "Neurodiversity may be every bit as crucial for the human race as biodiversity is for life in general."
>
> Harvey Blume

Why do we think differently?

We do not yet know the characteristic brain features that underlie most of these differences in mental functioning. However, a recent study found that people with six different neurodiverse conditions, including ASD, ADD/ADHD, and OCD (obsessive-compulsive disorder), all shared similar differences in the thickness of the cortex in some regions of the brain compared to neurotypical people.

DYSLEXIC

Dyslexic people have difficulty reading words, letters, or symbols. Dyslexia is thought to affect 15–20 percent of people.

NEUROTYPICAL

Anyone whose cognitive functioning is similar to the majority of people is referred to as "neurotypical."

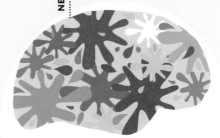

A personality disorder is a deep-set pattern of thoughts and behaviors developed by an individual that diverges greatly from the norm. These disorders are grouped according to whether these thoughts and behaviors are unusual (cluster A), dramatic (cluster B), or fear-based (cluster C). Personality disorders vary greatly. For example, borderline personality disorder (BPD) is marked by intense and changeable emotions and negative thought patterns. Avoidant personality disorder (APD) is defined by extreme social inhibition and self-imposed isolation. There is evidence that personality disorders are reflected in brain chemistry. For example, people with BPD show unusual brain activity in the amygdala, which regulates anxiety and anger, and the frontal lobe, which controls impulses.

DIFFICULT PATTERNS

CLUSTER A

ODD THINKING AND ECCENTRIC BEHAVIOR

Paranoid
Personality Disorder

Schizotypal
Personality Disorder

Schizoid
Personality Disorder

CLUSTER B

DRAMATIC AND ERRATIC BEHAVIOR

Antisocial
Personality Disorder

Narcissistic
Personality Disorder

Histrionic
Personality Disorder

Borderline
Personality Disorder

CLUSTER C

SEVERE ANXIETY AND FEAR

Avoidant
Personality Disorder

Dependent
Personality Disorder

Obsessive-compulsive
Personality Disorder

BRILLIANT MINDS

A person with exceptional intelligence or creative ability is often referred to as a genius. There are two kinds of genius—the first is exceptionally good at many things, while the second is profoundly talented in just one area. Both types have unusual brain architecture. People who are generally brilliant have an unusually large number of long neural connections that link different cognitive skills and give them more ways of looking at problems. They also have fewer dopamine receptors in the thalamus, meaning less information is inhibited from flowing to the cortex. As a result, more ideas reach consciousness. People who have remarkable ability in a narrow field have a larger than typical number of very short neural pathways in one or a few areas.

Differing brilliance

Brains that are creatively brilliant activate many parts to come up with ideas, while narrowly clever people use fewer but denser areas.

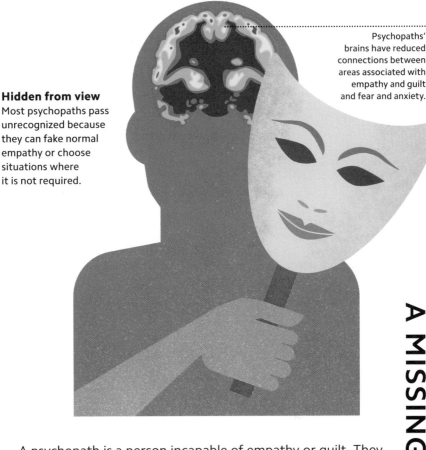

Hidden from view
Most psychopaths pass unrecognized because they can fake normal empathy or choose situations where it is not required.

Psychopaths' brains have reduced connections between areas associated with empathy and guilt and fear and anxiety.

A MISSING CONSCIENCE

A psychopath is a person incapable of empathy or guilt. They often display antisocial behaviors and can be impulsive. As a result, psychopaths, when unlikely to be caught, are inclined to commit criminal acts. Psychopaths' brains show less activation in frontal lobe areas that create feelings of empathy and morality. In neurotypical people, the sight of another in pain excites a network of neurons centered on the very front of the brain. This makes the viewer feel uncomfortable and might even produce pain. In psychopaths, this area does not become active when they see other people hurting. When prompted to imagine themselves in pain, however, that brain network becomes very active—more so than in neurotypical people.

A LOSS OF REALITY

Schizophrenia causes "positive" and "negative" symptoms. So-called positive symptoms include hallucinations (mostly visual and auditory), delusions (false beliefs that the majority do not share), and psychosis (loss of contact with reality). Negative symptoms include apathy, persistent fear, and at worst catatonia (a state of immobility). The exact causes of schizophrenia are unknown. It is thought to result from the triggering of a genetic predisposition to schizophrenia through environmental factors. The brain of a person with schizophrenia commonly contains excessive levels of dopamine, a neurotransmitter that passes signals within the brain. Typically, it also has enlarged lateral ventricles, exerting added pressure on the brain, and reduced cortical thickness, especially in frontotemporal areas. These structural features also occur in aging, leading some to believe that schizophrenia may represent accelerated aging.

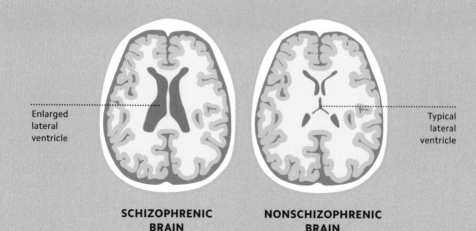

Enlarged
lateral
ventricle

Typical
lateral
ventricle

**SCHIZOPHRENIC
BRAIN**

**NONSCHIZOPHRENIC
BRAIN**

POSITIVE (PRESENT)

NEGATIVE (ABSENT)

DELUSIONS

Holding false beliefs even in the presence of contradictory evidence

FLATTENED AFFECT

Little interest or motivation in day-to-day activities, such as relationships

HALLUCINATIONS

Perceiving things that are not there, such as voices or objects

REDUCED SPEECH

Uneasy around others and with little motivation to add to conversations

DISORGANIZED SPEECH

Racing and out-of-control thoughts or ideas communicated incoherently

LACK OF INITIATIVE

Severe lack of motivation to reach end goals, such as paying bills or working at a job

Types of symptoms

Positive symptoms of schizophrenia are so-called because they represent thinking or behavior that was not present before and so can be viewed as having been added to the person's psyche. Negative symptoms, in contrast, represent the opposite—something having been taken away from the person.

A VARIED SPECTRUM

ASD is a lifelong developmental condition characterized by difficulties in social communication and over- or undersensitivity to sensory information, as well as a strong desire for routine and repetition. Autistic people fall on a spectrum. For example, many people with Asperger syndrome, which is just one type of autism, may have neurotypical language skills but greater social difficulties. Autistic traits usually appear in early childhood, and language difficulties are often a key indicator. Autism used to be explained as a lack of "theory of mind" (the ability to understand the mind states of others; see pp.104–105). The amygdala of autistic people is underactive when they try to decode the emotional facial expressions of others. However, new research shows that autistic people are more able to read the intentions of other autistic people, and neurotypical people can also struggle to read autistic people.

LESS AUTISTIC   **MORE AUTISTIC**

WHAT PEOPLE THINK THE AUTISM SPECTRUM LOOKS LIKE

"I am different, but not less."
Temple Grandin

A common misunderstanding
Autism varies greatly. One autistic person might face challenges that another autistic person never does or experience a symptom—such as fidgeting—to a much greater or lesser extent.

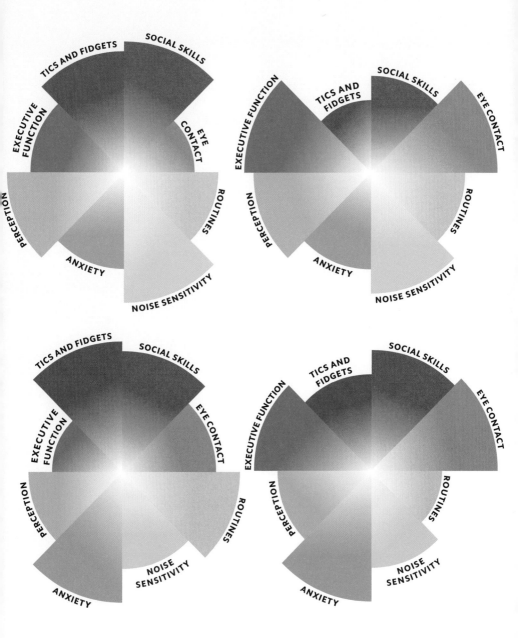

WHAT IT ACTUALLY LOOKS LIKE

THE

RATIO

BRAIN

N A L

Normal day-to-day existence calls on us to assess our situation moment by moment and to act appropriately. The basic drive to go forward, back off, or stay still comes from the unconscious emotional brain, but this part does not necessarily have the final say. Other parts of the brain may inhibit emotional impulses and moderate or even override them. Judgments and considered decisions are made by the brain's frontal lobes. They collate information gathered from the rest of the brain, including memories, and juggle this information to produce strategic action plans.

Not one of us
Sometimes, we intuitively prefer people similar to ourselves—even for job appointments, where we should be evaluating a candidate's potential objectively. This is called affinity bias.

INTUITION OR REASON?

When we say something is irrational, we mean that it seems illogical or contrary to evidence. However, much of our thinking is not ruled by logical reasoning. Our brains have evolved to use unconscious, intuitive thinking, which is much quicker than conscious deliberation and works well most of the time (although it is prone to biases). Some tasks, however—such as solving a complex arithmetical problem—do require logical thought, while other situations require both types of thinking. When buying a house, for example, it is rational to consider carefully the cost and other practical aspects. But there is also the intuitive, emotional factor: Do we actually want to live there? Ignoring this factor can lead to an expensive mistake, so rationality typically involves intuition as well as reasoning.

SPLIT-SECOND PROCESSING

At first, information coming in from the environment is unconscious and fragmentary. Visual information, for example, is initially registered as separate components—these include color, shape, position, and movement. These elements are then combined to form an image. In turn, this image is combined with associated sound, smell, emotion, words, and memories. All of this happens within half a second, which is the length of time it takes to become conscious of something. This process occurs as information is shunted forward along neural pathways in the brain. By the time you become conscious—for example, of a charging bull—it is an entire event.

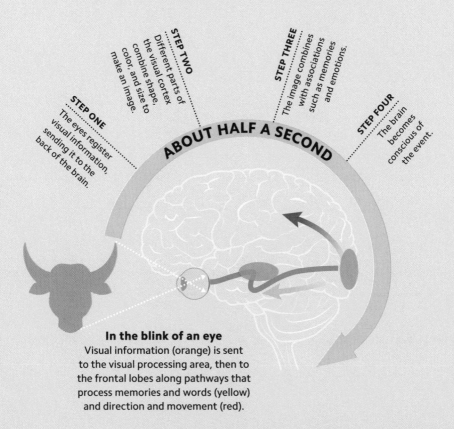

STEP TWO
Different parts of the visual cortex combine shape, color, and size to make an image.

STEP THREE
The image combines with associations such as memories and emotions.

STEP ONE
The eyes register visual information, sending it to the back of the brain.

STEP FOUR
The brain becomes conscious of the event.

ABOUT HALF A SECOND

In the blink of an eye
Visual information (orange) is sent to the visual processing area, then to the frontal lobes along pathways that process memories and words (yellow) and direction and movement (red).

WHAT GRABS YOU?

Most of our thoughts and memories are filtered out by the brain. Only those that provoke the most neural excitement reach consciousness. Some events grab our attention automatically because they are urgent or unusual, such as a car crash. For less sensational events, the brain must choose to attend. Some brains attend easily to bad events. Others distract from stimuli that could produce unpleasant emotions. Sometimes, our brain resists recalling something even though we experienced it consciously. This is called suppression.

Suppressed thoughts
Our brain can recall thoughts that were once suppressed if a similar event occurs that acts as a trigger. In this way, we can recover memories that were once set aside by our consciousness.

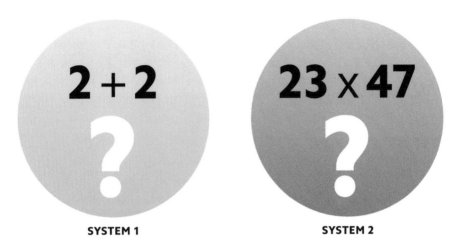

SYSTEM 1

SYSTEM 2

DOUBLE-EDGED THINKING

The brain has two systems for thinking: one fast and unconscious
(often called System 1) and the other slow, deliberate, and conscious
(System 2). Both systems feed into our decisions and judgments.
However, our conscious sense of decision-making is illusory: a "decision"
is just the conscious awareness of what our bodies are about to do. If
we decide to make a physical movement, such as moving to hit a tennis
ball, our awareness of it happens only after our brain has sent out
the orders to our muscles. The brain then effectively back-dates the
sense of deciding to before the action. After acting, we often use our
conscious mind to rationalize the decision and to provide justifications
for making it that may or may not actually apply.

ISSUING ORDERS

To produce a considered physical action, the brain's frontal lobes need to plan, coordinate, and execute movements. In the frontal lobes, it is the job of the supplementary motor cortex to work out the sequence of nerve impulses to send to the motor cortex in order to achieve a desired movement. During sleep, the brain rehearses these sequences, helping us consolidate learning and create automatic shortcuts to use later. A neurological disorder such as focal hand dystonia can interrupt these processes. This can develop after extensive piano practice and prevents players from moving certain fingers separately. This is because the motor areas in the brain controlling those digits have fused together.

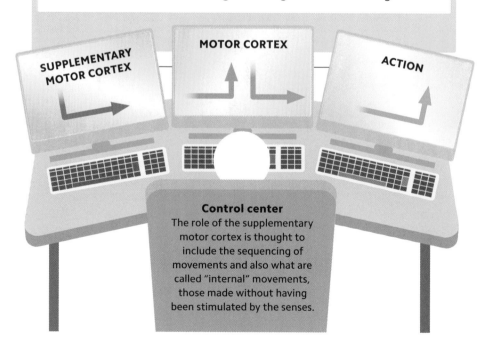

SUPPLEMENTARY MOTOR CORTEX

MOTOR CORTEX

ACTION

Control center
The role of the supplementary motor cortex is thought to include the sequencing of movements and also what are called "internal" movements, those made without having been stimulated by the senses.

Caught in the middle
Curbing our impulses can lead to delayed gratification. However, research has shown that too much self-restraint can also lead us to take less pleasure in our own achievements.

IMMEDIATE TEMPTATION

Immediate rewards are often tempting to choose, even at the expense of long-term goals. A recurring and compulsive need for quick gratification is a trait of addiction.

Studies show that close social relationships help us achieve our goals. A good night's sleep can also help us keep our temptations in check.

LONG-TERM GOAL

CHOOSING A PATH

The frontal lobes "decide" on a goal and keep the brain on track by inhibiting any competing impulses that may arise in the limbic system (see p.16). For example, we may decide to eat a single piece of chocolate but find ourselves seized by the temptation to eat more. If frontal lobe activity is strong, it can overcome such temptation and hold the course that achieves long-term satisfaction. Decision-making involves weighing available options and imagining the consequences of each. The process may be conscious, such as deciding where to go on vacation, or unconscious, such as deciding exactly where to place our feet when dancing.

THE BIRTH OF IDEAS

Coming up with creative ideas and insights involves using existing knowledge in new ways. Doing so is a whole-brain activity that involves shutting down parts of the prefrontal cortex involved in self-monitoring, self-criticism, and inhibition. People with frontal dementia may struggle to regulate their thoughts but may also experience a spontaneous artistic creativity that they lacked before illness. Creativity is also associated with increased activity in areas in the default mode network (see pp.50–51) that are associated with daydreaming.

Stimulating new thoughts
Using noninvasive brain stimulation to influence neurons in areas associated with creativity can lead people to solve problems they couldn't tackle beforehand and inspire them to find more creative uses for everyday objects.

Your brain on creativity

When we are creative, we tend to focus on our inner world and switch on parts of the brain involved in self-expression and creating an autobiographical narrative. This includes the medial prefrontal cortex and the default mode network.

A NEW STATE OF MIND

Hypnosis occurs when a person shifts into a state of mind in which they are unusually compliant, uninhibited, or capable of intense imagination. Under hypnosis, people may be told, or choose to imagine, that they see things differently than usual. For example, they might see red when looking at a yellow object, and this is reflected in brain function—the neurons in their brain that react to red become active, while those concerned with yellow do not.

QUIETING THE BRAIN

Meditation
is a practice in which a
person seeks a higher sense of
awareness, attention, and calm. There
are many different types of meditation, such
as mindfulness, which involves noticing our
thoughts and experiences with curiosity rather
than judgment, and Zen, which is designed to induce
a no-thinking state. Meditation can affect our brains
and induce an altered state (see p.147). During
meditation, the amygdala—the seat of feelings such
as fear, anxiety, and panic—calms down. If meditation
is practiced regularly over a long period of time, it
can help control stress; reduce rumination; and
improve resilience, attention, and
emotional regulation.

THE
CONSC
BRAIN

IOUS

One of the most extraordinary things about the universe is that parts of it are conscious. No one can be sure which parts because consciousness is private. We assume, though, that they include living people and some other creatures. No one knows what consciousness is, nor what, if anything, it is for. Philosophers argue about whether it is generated by brain activity, a side effect of it, the physical process itself, or nothing whatever to do with brains. There are even philosophers who deny that consciousness exists at all. Scientists approach the mystery by trying to determine what brain activity occurs in a person when the person reports a particular experience.

DUALISM

This is the idea that consciousness is "spiritual stuff" instead of brain activity or its product.

WHAT IS CONSCIOUSNESS?

MONISM

This theory states that brain activity itself is consciousness.

VISUAL DATA

Nobody knows what consciousness is, even though philosophers have grappled with the question for millennia. Scientists have identified brain processes that are linked to consciousness, the so-called neural correlates of consciousness (see pp.144–145), but these do not explain what it is. Theories of consciousness include dualism, in which it is seen as separate from the material world and its laws; monism, which states it is neuronal activity; emergence, the idea that it emerges from (but is not identical to) neural activity; and property dualism, in which it is an aspect of matter that is not yet understood.

CONSCIOUS

Levels of consciousness
Freud proposed that the mind comprises three levels: the conscious mind, containing the ideas and thoughts that we are currently aware of; the preconscious, which stores information and memories that are easily recalled; and the unconscious, which holds suppressed desires, memories, and instincts only accessible by psychoanalysis.

PRECONSCIOUS

"The unconscious is ...
as much unknown to us as the
reality of the external world."
Sigmund Freud

UNCONSCIOUS

UNLOCK THE UNCONSCIOUS

Our brains are active all the time, even during deep sleep, but we are conscious of very little of this. Unconscious activity includes instinctive reflexes, such as breathing, blinking, and constant small muscle adjustments to maintain balance, as well as most emotions. The unconscious mind can execute complex actions: many people have driven a familiar route and only realized later that they have no recollection of doing so. Psychoanalyst Sigmund Freud popularized the notion of the unconscious mind and theorized that some thoughts and desires are too disturbing for a person to acknowledge consciously. He devised psychoanalysis to help access these suppressed thoughts. Much of Freudian theory has been discredited, but his basic structure remains useful.

SYNCHRONOUS FIRING

Strong evidence of a link between consciousness and synchronous activity in some neuronal groups has been identified.

HIGH FIRING RATES

High-speed oscillations in neurons, which equate to more than 40 on/off cycles per second, are associated with conscious thought.

FRONTAL ACTIVITY

Brain-imaging techniques have identified widespread cortical activity in the frontal lobes as a neural correlate of consciousness.

> "Everything the human brain does is amazing but not all of it special. It is only the rich and highly developed quality of our consciousness that distinguishes us from other species."
>
> Rita Carter

SENSORY INPUT

FINDING CONSCIOUSNESS

Scientific studies of the brain have discovered several different types of neural activity that mark a state of consciousness in an individual. These so-called "neural correlates of consciousness" (NCC) are the minimal neural mechanisms required in order to generate any one specific conscious experience. In other words, a person's neural state has been shown to correlate directly to their conscious state. Brain-imaging techniques (see pp.26–27) have revealed that brain activity varies according to the particular experience, but the neural correlates of consciousness seem to be necessary for there to be any type of consciousness.

ARTIFICIAL INTELLIGENCE

Machines can perform many tasks that people carry out, often with better results. For example, computers are much faster and more accurate at calculations than us. However, the human brain has certain features that are impossible to replicate in disembodied machines because they rely on the brain's constant interplay with the rest of the body. The closest reproduction of human intelligence in a machine is virtual neural networks. Cutting-edge artificial intelligence (AI) systems mimic human intelligence by feeding back the results of their actions and adapting their behavior accordingly. These "self-learning" machines can beat humans at complex games, but whether a virtual brain is or will ever be conscious is unknown.

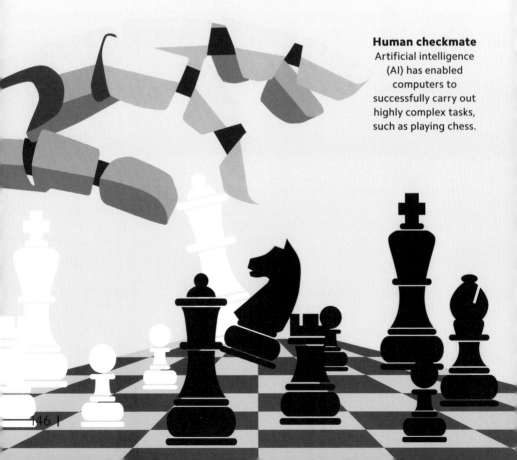

Human checkmate
Artificial intelligence (AI) has enabled computers to successfully carry out highly complex tasks, such as playing chess.

Psychological
Certain practices, including meditation, trances, sensory deprivation, and hypnosis, can create an altered mind-state.

Pharmacological
Drugs, such as alcohol, cannabis, or opioids, can disrupt how brain chemicals function and alter both the user's awareness and their levels of consciousness.

Physical
Experiencing extreme environmental conditions, such as being at high altitudes, can induce altered states, as can long periods of fasting.

Spontaneous
Spontaneously induced altered states include daydreaming, near-death experiences, and the hypnagogic state, which occurs just as we fall asleep.

Disease-induced
Illness and disease can affect conscious experience, causing psychotic disorders, epileptic seizures, and comas.

FRAMES OF CONSCIOUSNESS

Human brain functioning evolved to keep us safe, and in most circumstances it does not produce extreme emotions or extend our thoughts or senses beyond a fairly narrow range. However, certain conditions, such as extreme situations, the influence of drugs, brain injuries, and some self-determined practices (including meditation), can nudge us out of this mind-set. The strange mind-states that our brains can produce may be ecstatic or nightmarish, momentary, or long-lasting. Some may give a worldview that seems to extend beyond the material, while others produce sensory distortions or a totally new perspective on the world.

TIME AND THE BRAIN

Biological clocks exist in every cell of our body, and the brain has several ways of producing a sense of time passing. One area of the brain responsible for keeping the body to a roughly 24-hour rhythm is the suprachiasmatic nucleus (a tiny nucleus above the hypothalamus), also known as the brain's central timekeeper. In addition, a neural circuit that passes through the frontal lobe plays a further role in measuring the amount of time that passes. The circuit achieves this by registering the number of events—spikes in brain activity—that occur within each "cycle" of neural activity.

DOPAMINE PRODUCTION

The basal ganglia is a cluster of structures that includes the substantia nigra, which contains neurons that produce dopamine.

NEURONAL CIRCUIT

Dopamine flows from the basal ganglia to the prefrontal cortex in regular cycles. Each cycle represents a perceived "packet" of time.

Body clock

Dopamine, the primary neurotransmitter responsible for time processing, flows between the substantia nigra, the basal ganglia, and the prefrontal cortex in order to control the circadian rhythm (see p.150).

PACEMAKER

The suprachiasmatic nucleus is located above the hypothalamus and is responsible for regulating most circadian rhythms in the body, acting as a pacemaker for our internal timing systems.

Since the year 2000, the average attention span dropped from 12 seconds to 8 seconds.

GRABBING YOUR ATTENTION

Beyond control
Many of the brain's mechanisms that direct our attention are not under conscious control. For example, it is impossible not to attend to a sudden jolt.

Attention is like a spotlight—it focuses on particular aspects of available experience and blocks out the rest. For example, when reading a book in public, our brain directs our senses to the text, even when there are distractions. It is possible to consciously direct attention, perhaps by looking intently at an object. Our attention system is spread across the brain, so any damage may affect concentration on a particular task. Stroke damage to one hemisphere may result in difficulty focusing on one side of the visual field, even when there is no sensory loss.

NECESSARY REST

Although scientists have yet to discover why sleep is so important, the daily switch-off from normal consciousness is an essential part of our circadian rhythm (our body's 24-hour cycle). If we don't get enough sleep, we suffer cognitive dysfunction (such as memory loss), and chronic deprivation is associated with heart disease, obesity, and mood disorders. The brain remains active, even in the deepest stages of sleep, when memories are consolidated.

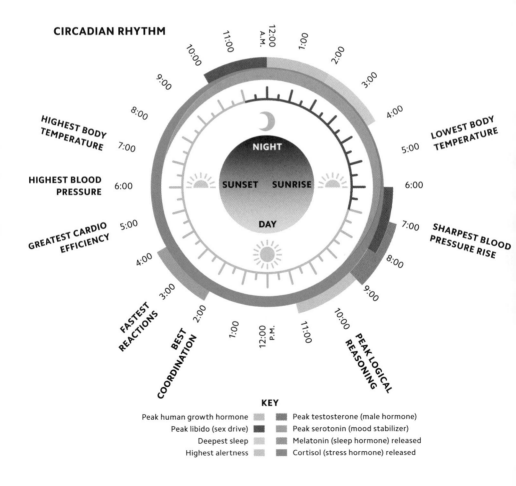

CIRCADIAN RHYTHM

KEY

Peak human growth hormone	Peak testosterone (male hormone)
Peak libido (sex drive)	Peak serotonin (mood stabilizer)
Deepest sleep	Melatonin (sleep hormone) released
Highest alertness	Cortisol (stress hormone) released

DEEP SLEEP — The longest periods of deep sleep occur at the start of the night.

REM — Dream sleep is called REM (Rapid Eye Movement) sleep because our eyes dart around when dreaming.

2:00
3:00
4:00
1:00
5:00
12:00
6:00
11:00
7:00

AWAKE
REM
LEVEL 1
LEVEL 2
LEVEL 3

LIGHT SLEEP
DEEP SLEEP

AWAKE — Neurons switch on and off 12–40 times per second when we're awake.

LEVEL 2 — In deeper sleep stages, neurons switch on and off as few as 3 times a second.

TYPES OF SLEEP

During a normal night's sleep, a person cycles through four stages of sleep. Each cycle last about 90 minutes. In the early stages, the brain gradually slows down, but the sleeper can still be roused easily. In stages two and three, brain activity is even lower. Deep sleep is thought to be the most restorative. After 20 to 30 minutes in this state, the sleeper's brain becomes more active, and then it generates dreams (see p.153).

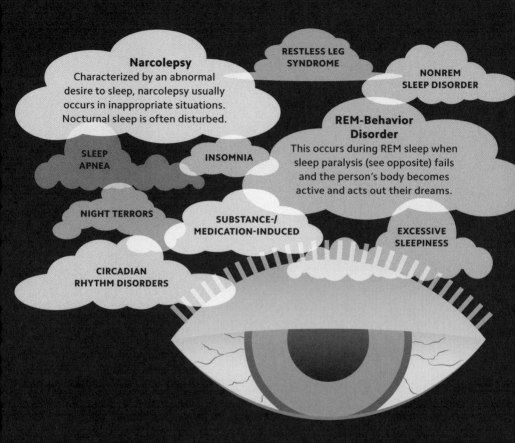

Narcolepsy
Characterized by an abnormal desire to sleep, narcolepsy usually occurs in inappropriate situations. Nocturnal sleep is often disturbed.

RESTLESS LEG SYNDROME

NONREM SLEEP DISORDER

SLEEP APNEA

INSOMNIA

REM-Behavior Disorder
This occurs during REM sleep when sleep paralysis (see opposite) fails and the person's body becomes active and acts out their dreams.

NIGHT TERRORS

SUBSTANCE-/ MEDICATION-INDUCED

EXCESSIVE SLEEPINESS

CIRCADIAN RHYTHM DISORDERS

STRUGGLES WITH SLEEP

A sleep disorder is a condition that routinely affects the quality of a person's sleep and impacts daily life. Secondary sleep disorders result from external conditions, such as environmental disturbances, anxiety, or depression. If these are managed, sleep often improves. Primary sleep disorders have no external trigger. The most common is insomnia: the inability to fall (or stay) asleep. Adopting regular, disciplined sleep habits may cure the problem, but severe cases may require drugs to enhance the effects of anti-excitatory chemicals in the brain.

Strange simulations
When we dream, instead of drawing on external sensations, the brain generates the dream world itself by pulling up a mixture of fragmented memories and thoughts and knitting them together to present a bizarre simulation of the waking world.

DREAM WORLDS

Three or four times each night, most of us enter the mind-state known as dreaming. Within a few minutes, our brain shifts from a state of very low activity (deep sleep) to a partial state of high arousal. The parts of the brain that become active are those at the back and top, where sensory information is generated and woven into perception. We are often very active in our dreams, but sleep paralysis—which blocks motor commands from the brain to the muscles—prevents us from acting them out.

THE DREAMING MIND

Dreams easily slip from our waking minds, and many people recall nothing of them at all. This is thought to be because the brain's hippocampus—the nucleus that encodes experiences as a first step toward laying them down in memory—works differently during dreams, linking newly registered experiences with existing ones instead of encoding current ones. The typical irrationality of dreams is due to inactivity in parts of the frontal lobes that normally act as a "reality-tester." Without this, our brains can generate experiences during dreams that would be impossible in the real world.

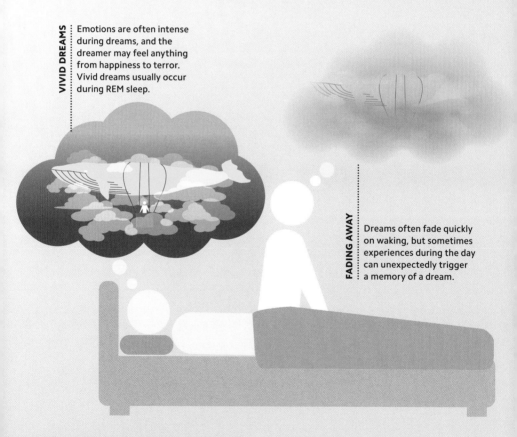

VIVID DREAMS
Emotions are often intense during dreams, and the dreamer may feel anything from happiness to terror. Vivid dreams usually occur during REM sleep.

FADING AWAY
Dreams often fade quickly on waking, but sometimes experiences during the day can unexpectedly trigger a memory of a dream.

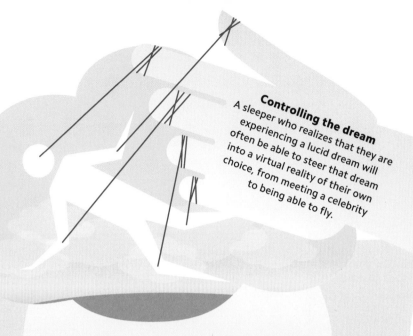

Controlling the dream
A sleeper who realizes that they are experiencing a lucid dream will often be able to steer that dream into a virtual reality of their own choice, from meeting a celebrity to being able to fly.

FALSE AWAKENING

A type of dream consciousness known as lucid dreaming occurs when the brain continues to create a dreamscape, in all its intensity and wildness, but the sleeper feels as though they have woken up. What happens is that the reality-checking brain mechanism (see opposite) kicks in even as the dreamscape continues to unfold. However, unlike with ordinary waking, during lucid dreaming, the sleeper remains paralyzed and unable to sense the outside world. A person who has a lucid dream without realizing it may find it very frightening as they fight unsuccessfully to move while being trapped within their hallucination.

INDEX

Page numbers in **bold** refer to main entries.

A

action 10, 46, 89, 133, 134
  appropriate 47, 129
  conscious 93
  planning **94–95**
action potentials 22
adaptability **18–19**
addiction 77
adolescence 45, 58
adrenaline 49
adult brain 59
affinity bias 130
aging 59, 124
alcohol 147
alpha waves 24
altered states 138, 139,
  147
altruism 109
Alzheimer's disease 113
amputations 91
amygdala 16, 41, 46, 47, 48,
  70, 84, 120, 126, 139
anger 49, 120
anxiety 25, 77, 120, 121,
  139
apathy 124, 125
art, visual **50–51**
articulation 82, 86
artificial intelligence **146**
artistic creativity 136
Asperger syndrome 126
attachment 114
attention 86, 87, 94, 132,
  139, **149**
attention deficit disorder/
  attention deficit
  hyperactivity disorder
  (ADD/ADHD) 117, 118,
  119

attraction 114
auditory cortex 87
autism spectrum 104, 117,
  118, 119, 122, **126–127**
avoidant personality
  disorder (APD) 120
awareness 46
  higher sense of 139
axons 9, 13, 18, 20, 21, 22,
  23

B

babies
  language 80, 81
  mind reading 106
background noise 35
balance 93, 99
basal ganglia 70, 77
behavior 11, 41, 44, 63, 67,
  117
  habits **76–77**
  regulating 135
belonging, sense of 110
beta waves 24
bias 110
bipolar disorder 45
birth 58
blood flow 26
body, left and right side 12,
  13
body clock **148**
body dysmorphic disorder
  (BDD) 92
body language **84–85**
body maps **90**
bonding **109**
borderline personality
  disorder (BPD) 120
brain waves **24**, 25

brainstem 36, 93
breastfeeding 109
Broca's area 82

C

catatonia 124
central nervous system 96
cerebellum 14, 15, 70, 93, 99
cerebral cortex 14, **15**, 70,
  73, 98, 99
chemoception 29
childbirth 109
childhood 58, 62
circadian rhythm 148, 150
cocktail party effect 35
cognitive network **93**
cognitive skills 61
colors 32, 39
communication 79–87
computers 146
connectivity 60
conscious action 93
conscious decisions 89, 133
conscious memories 69
consciousness 141, **142**, 143
  dream **154**
  of events 131
  and filtered memories 132
  and rationality 130
  science of **144–145**
  and thinking 133
consensus 110
control **135**
coordination 93, 94, **98–99**
COVID-19 113
creativity **136–137**
criminality 123
cues/triggers 77, 94
culture 60, 63

D

danger 16
daydreaming 51, 136, 147
decision-making 95, 129, **133**
declarative memories 69
deep sleep 151
default mode network (DMN) 50, 51, 110, 136, 137
delta waves 22
delusions 124
dementia 87, 136
dendrites 18, 19, 20, 21, 24,
depression 41, 45
development, brain **58–59**, 117
disease
  altered states 147
  recovering from 24, 25
DNA 56, 57
dopamine 48, 109, 114, 122, 124, 135, 148
dreams 151, **153**
  dream consciousness **154**
  lucid dreaming **155**
drugs 147
dualism 142
dyscalculia 117
dyslexia 117, 118, 119
dysmorphia **92**
dyspraxia 118

E

ears 30, 31, 34, 35
EEG (electroencephalogram) 22, 23
electrical activity 22, 27
electrical signals 9, 19, **20–21**, 61
emotional range **43**
emotional response **42**

emotions 9, 11, 15, 16, 41, 67, 130, 131, 154
  and brain **46**
  and memory 70
  positive and negative **48–49**
  reading 126
  regulating 139
  unconscious **47**
empathy 101, 104, **108**, 109, 110, 112, 117, 123
environment 41, **56–57**, 60, 61, 89, 124, 147, 152
epigenetics 57
epinephrine 42
evidence, gathering **131**
evolution 10
experience 9, 55, 56–57, 58, 59, **62**, 67, 141
  conscious **144–145**
explicit memory 70
extroversion 63
extroverts 64, 65
eyes 30, 32, 61

F

facial expressions 43, 47, 79, 84, 104, 126
facial recognition 72, **102–103**
false memory 67, **75**
fasting 147
fear 49, 124, 139
female brain **60**
fidgeting 126
fight or flight 42
fine motor skills **98–99**
fluid intelligence 59
fMRI (functional MRI) scans 26
focal hand dystonia 134
focus 94, **149**
forebrain 14
forgetting **73**
Freud, Sigmund 143

frontal lobe 15, 82, 93, 99, 123, 129, 134, 135, 148, 149, 154
fusiform face area 102
fusiform gyrus 72, 102

G

gamma waves 22
gender **115**
  differences 60
genetics 9, 55, **56–57**
  and the brain **61**, 62
  and neurodiversity 124
  and personality 63
genius 117, **122**
gestures 79, 81, 83, 84
goals, long-term 135
gray matter 10
gross motor skills **98–99**
guilt 112, 123

H

habits **76–77**
hallucinations **33**, 124, 125, 155
hands, left and right 12, 13
happiness 56
health, and sleep 150
hearing 29, 30, 31, **34–35**, 47
hemispheres, cerebral 9, **12–13**, 79
  male and female brains 60
herd instinct 101, **110–111**
hindbrain 14
hippocampus 16, 70, 73, 74, 120, 154
homunculus, sensory 36
hormones 16, 29, 42, 49, 108, 109, 114
human brain **10–11**
hunger 16, 17
hyperactivity 117
hypnagogic state 147
hypnosis **138**, 147
hypothalamus 16

I

images 131
imagination 10, 33, 50, 138
imaging, brain **26–27**, 144, 145
imitation 107
implicit memory 70
impulses 120, 135
in-group 110
individuality 55, 56, **64–65**
information processing 10, 11, 15, 131
inheritance 56, 61, 62
injuries, brain 147
insomnia 152
instincts 11, 143
intelligence 15, 61, 122
artificial **146**
interoception 29
internal representation 91
introverts/introversion 63, 64, 65
intuition 130
ion channels 20, 21
isolation, social 110

J

joints 96, 97
judgments **133**
Jung, Carl 65
justifications 133

K

kinesthesia **96–97**
knowledge 59

L

language 12, 13, 15, 69
hardwired **80**
understanding **81**
lateral ventricles 124
leading questions 75
learning 24, 62, 67, 70, 134

left hemisphere 12, 13, 46, 86
lens 32
light 30, 32
limbic system **16**, 41, 46, 109, 135
listening 83
literacy/illiteracy **86–87**
lobes 9, **15**
logic 130
loneliness **113**
long-term memory 68–69
love **114**
lust 114

M

magnetoencephalography (MEG) **27**
male brain **60**
mania 41, 45
meaning 86, 87
meditation **139**, 147
memory 11, 15, 16, 24, 67, 129, 131
aids **74**
categories **68–69**
distortions **75**
dreams 154
editing the facts **132**
habits **76–77**
loss of **73**
recognition **72**
recovered 132
and sleep 150
and smell 38
storage and recall **70–71**
mental disorders 55
microvilli 37
midbrain 14
mime 81
mind reading **106**
mindfulness 139
mirror neurons **107**, 108
mnemonics 74
monism 142
moods 41, **44–45**, 63
morality 101, **112**

motor cortex 89, 90, 93, 94, 95, 134
motor skills, gross and fine **98–99**
movement 10, 15, 32, 89, 96, **98–99**, 133, 134
memory sequences 94
MRI (magnetic resonance imaging) scans 26
muscle dysmorphia 92
muscles 89, 96, 97, 133
music 39
Myers–Briggs model 64–65

N

narcolepsy 152
native language 81
natural selection 10
nature vs. nurture **56–57**
negative emotions 49
neural activity **22**, 27, 144, 145, 148
neural correlates of consciousness (NCC) 142, 145
neural networks 18, 70–71
neural pathways 9, 18, 19, 24–25, 39, 55, 61, 62, 71, 122, 131
neural plasticity **24–25**
neurodiversity 117, **118–119**
neurofeedback **23**
neurons 9, **18**,19, 20–21, 22, 58, 91
mirror **107**
senses 30
neurotransmitters 19, 20, 48, 124, 148
neurotypical 118, 119
nondeclarative memories 69
nonverbal communication **83–85**
noradrenaline 49
nose 38

O

object recognition pathway 72
observation 107
obsessive-compulsive disorder (OCD) 77, 119
occipital lobe 15, 93
olfactory cortex 37, 38
optic nerve 32
orbitofrontal cortex 38, 135
ossicles 34
oxytocin 108, 109, 111, 114

P

pain 36, 108
panic 139
papillae 37
parietal lobe 15, 93, 149
Pavlov, Ivan 68
perceptions 39, 44
  distorted 92
  false **33**
  visual 32
personality 15, **63**, **64–65**
personality disorders **120–121**
perspective, other people's **104–105**
phantom limbs **91**
philosophy 141
physiological needs 17
pitch 34
planning 58
plasticity 62
pleasure 48
position, sense of 96–97
positive emotions **48**, 109
postsynaptic neurons 19
preconscious 95, 143
precuneus 106
prediction 89
prefrontal cortex 58, 106, 109, 114, 136, 137, 148
premotor areas 93
primary sense areas 29
problem-solving 136
pronunciation 82
property dualism 142
proprioception 29, **96–97**
prosopagnosia 102
psychopaths 104, **123**
psychosis 124, 147
psychotherapy 143
pupils 84, 85

R

rationality **130**
reading **86–87**
reality checks 154, 155
reasoning 58
  abstract 12
recall, memory 33, **71**, 73
receptor cells 30, 38
recognition **72**
reflection **50–51**
reflexes 58, 143
regions, brain **14**
REM (rapid eye movement) sleep 151, 152, 154
repetition 126
resting potential 20
retina 30, 32
reward pathways 77, 114
rewards 46, 77, 135
rhythm 79
right and wrong 112
right hemisphere 12, 46
romantic attachment 109
routine 77, 126
rumination 139

S

safety **17**
Sally-Anne Test 105
scanning, brain **26**, 113
schizophrenia **124–125**
self-control 58, 135
self-criticism 136
self-expression 137
self-monitoring 136
self-restraint 135
sensations 9
senses 10, 11, 12, 29, **30–39**, 93
  gathering evidence 131
  synesthesia **39**
sensory deprivation 147
sensory information 94, 95
sensory perception 93
serotonin 42, 48, 109
sexual arousal 109
sexuality **115**
shame 112
short-term memory 68
sign language 81, **83**
signals 95
size 10, 60
skills 59, 94
skin 36
sleep 70–71, 74, 134, 135, **150–151**
  disorders **152**
  dreaming **153–155**
smell 16, 29, 30, 37, **38**, 47
social anxiety 109
social bonding 109
social brain **112–113**
social skills 101
sociopaths 123
somatosensory cortex 90, 91
sound, filtering **35**
sound waves 34
speaking **82–83**
speech 81, 82, 83, 86, 87, 125
spinal cord 9, 36, 89
spindle fibers 97
spine 95
stimuli 29, 36, 47
storage systems 10, 68, **70**, 73
strategic action plans 129
stress 49, 109, 113, 139
striatum 76, 77, 93
structural plasticity 24
substance abuse 147

superconducting quantum interference devices (SQUIDs) 27
supplementary motor cortex 134
suppression 132, 143
suprachiasmatic nucleus 148
sympathy 108
synapses 18, **19**, 20, 24
synaptic clefts 19
synaptic plasticity 24
synesthesia **39**

T

taste 29, 30, 31, **37**
temperature regulation 16, 17
temporal lobe 15
temporal-parietal junction 106
tendons 97
testosterone 42, 60, 115
thalamus 32, 122
theory of mind 101, **104–105**, 106, 126
theta waves 22
thirst 17

thoughts 9, 10, 44, 46, 93
conscious and unconscious **133**
stimulating new 136
threats 46
time **148**
tonal expression 79
tone 34
touch 29, 30, 31, **36**, 91
traits, personality 63
trances 147
Transcendental Meditation 139
trauma, recovery from 25
trolley problem 112
trust 110, 113
twins
identical 55, 56, 62
nonidentical 56

U

unconscious memories 69
unconscious thought 133
unconsciousness 130, 131, **143**
understanding **81**, 82, 83
other people **104–105**, 106
urges 77

V

verbal communication **82–83**
vibrations 31, 34
virtual reality 155
vision 29, 30, **32**, 47, 93
visual cortex 32, 61, 72, 131
visual information 131
visualization 74
visuospatial attention 12
vocalizations 80
volume 10
voluntary movement 93

W

weight 10
Wernicke's area 81, 82, 83
white matter 10
working memory 68
writing **86–87**

Z

Zen 139

ACKNOWLEDGMENTS

DK would like to thank the following for their help with this book: Ali Scrivens for illustrations; Steve Bere and Laura Gardner for help with design; Katie John for proofreading; Helen Peters for the index; Senior Jacket Designer Suhita Dharamjit; Senior DTP Designer Harish Aggarwal; Jackets Editorial Coordinator Priyanka Sharma; Managing Jackets Editor Saloni Singh.

All images © Dorling Kindersley
For further information see:
www.dkimages.com

Quotations:
45: Carrie Flsher, *Psychology Today*, Nov/Dec 2001.
59: Tim Berners-Lee, *Weaving the Web*, 1999
68: George Johnson, *In the Palaces of Memory*, 1991
102: Oliver Sacks, *The New Yorker*, August 23, 2010
119: Harvey Blume, *The Atlantic*, September 1998
126: Temple Grandin.

SIMPLY EXPLAINED

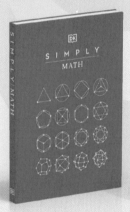

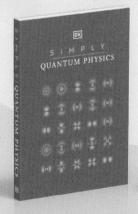

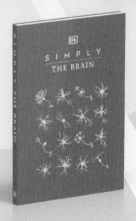

DK · For the curious